Environmental Justice for Climate Refugees

This book explores who the climate refugees are and how environmental justice might be used to overcome legal obstacles preventing them from being recognized at an international level.

Francesca Rosignoli begins by exploring the conceptual and complex issues that surround the very existence of climate refugees and investigates the magnitude of the phenomenon in its current and future estimates. Reframing the debate using an environment justice perspective, she examines who has the responsibility of assisting climate refugees (state versus non-state actors), the various legal solutions available and the political scenarios that should be advanced in order to govern this issue in the long term. Overall, Environmental Justice for Climate Refugees presents a critical interrogation of how this specific strand of forced migration is currently categorized by existing legal, ethical, and political definitions, and highlights the importance of applying a justice perspective to this issue.

Exploring the phenomenon of climate refugees through a multi-disciplinary lens, this book will be of great interest to students and scholars of environmental migration and displacement, environmental politics and governance, and refugee studies.

Francesca Rosignoli is Junior Global Horizons fellow in the Swedish Collegium for Advanced Study (SCAS) in Uppsala, Sweden.

T0210709

Routledge Studies in Environmental Migration, Displacement and Resettlement

For more information about this series, please visit: www.routledge.com/Routledge-Studies-in-Environmental-Migration-Displacement-and-Resettlement/book-series/RSEMDR

Environmental Justice for Climate Refugees

Francesca Rosignoli

Routledge
Taylor & Francis Group

NEW YORK AND LONDON

from Routledge

First published 2022
by Routledge
4 Park Square, Milton Park, Abingdon, Oxon OX14 4RN

and by Routledge
605 Third Avenue, New York, NY 10158

Routledge is an imprint of the Taylor & Francis Group, an informa business

© 2022 Francesca Rosignoli

British Library Cataloguing-in-Publication Data
A catalogue record for this book is available from the British Library

Library of Congress Cataloging-in-Publication Data
A catalog record for this book has been requested

ISBN: 978-0-367-60943-6 (hbk)
ISBN: 978-0-367-60945-0 (pbk)
ISBN: 978-1-003-10263-2 (ebk)

DOI: 10.4324/9781003102632

Typeset in Times New Roman
by Apex CoVantage, LLC

To Roberto

Chinese, gypsies, blacks they all feel blue
Migrants die in the deep sea through
The teeth of sharks that have no name
They lose their hopes and their long, long way
I have no voice in this beautiful jail
I have no chance in this umpteenth rape
(Rosignoli, 2017)

Contents

Acknowledgments

This book is the result of long and careful research started at Stockholm University in February 2019 and finalized at The Swedish Collegium for Advanced Study (SCAS), Uppsala, in November 2021, where I was a Junior Global Horizons Fellow funded by the Stiftelsen Riksbankens Jubileumsfond (RJ). I gratefully acknowledge both Swedish Institutes for their financial support of this research. In particular, I want to express my gratitude to the professors who have been following my work: Professors Karin Bäckstrand, Jonas Ebbesson, and Kerstin Lidén of Stockholm University, and Professors Marco Armiero and Sverker Sörlin of Tekniska högskolan (KTH). I also want to thank Professor Avidan Kent for his interest and encouragement in my research. A very special thanks to Professor Michael Watts, Senior Global Horizons Fellow at SCAS, for revising the first draft of my manuscript and his precious suggestions.

1 "Climate Refugees"

Toward the construction of a new subjectivity

Introduction

What is a "climate refugee"? Is this the right term to use? Are "climate refugees" an identifiable and "measurable" group of people? Even today, the answers to these questions remain controversial, mostly because of the heterogeneous and multicausal nature of climate-induced migration. Notably, evidence shows that different types of environmental degradation are likely to trigger different patterns of migration (Kälin & Schrepfer, 2012). To begin with, a first divide is between slow-onset and rapid-onset events.

Slow-onset events are characterized by adverse long-term impacts on the environment, such as sea-level rise, groundwater and soil salinization, droughts, and desertification. Options available to people affected by slow-onset events depend upon the stage of environmental degradation, ranging from staying, migrating or eventually escaping when compelled to. These strategies can be permanent or temporary, depending on circumstance and context. For example, Small Island Developing States (SIDS) represent a particular case of slow-onset disasters in the later stages of environmental degradation making them uninhabitable, the *last-resort* of planned relocations are becoming ever more frequent (Ionesco, 2017). By contrast, rapid-onset events such as floods, storms, hurricanes, typhoons, cyclones, and mudslides, all have negative short-term impacts making large-scale displacement the most common consequence. The immediacy of impact makes evacuation the only available option, this movement is however often temporary with a return made as soon as conditions allow.

Despite differences in climate migration patterns, there is a common denominator: people fleeing the impacts of environmental disruptions rarely cross state borders when migrating (Rigaud et al., 2018) and move within their country of origin as Internally displaced persons (IDPs) (Ionesco, 2017; Rigaud et al., 2018).[1] A further common element relies upon the multicausal nature of climate-induced migration. Environmental factors can and do co-exist with others such as unemployment, war, violence, and social networks in countries of destination, which work together as drivers for migration. Understanding how and to what extent these factors work together is one of the most pressing challenges to understanding factors influencing human mobility (McLeman & Gemenne, 2018).

DOI: 10.4324/9781003102632-1

To sum up, the multi-causal and heterogeneous nature of climate-induced migration implies that it must be viewed through the lens of space, time, and context. Not only do these factors offer a key to understanding migration but they also indicate the extent and likelihood of the affected population to survive, and therefore their legal status. Changes in people's legal status are ultimately determined by municipalities adopting legal categories, the main questions are: Does the existing legal framework have the necessary tools to cope with a multi-causal and heterogenous phenomenon such as climate-induced migration? Is there a way out of the "legal impasse" that prevents from recognizing "climate refugees" as a new (legal) subjectivity? The chapter seeks to answer these questions by a genealogical examination of empirical controversies, terminological disputes, and struggles of major players competing for this emerging area of policymaking.

Methodology

This analysis is conducted through the genealogical method. In particular, I draw on the Foucauldian concept of subjugated knowledges conceived in its twofold understanding of historical contents and disqualified knowledges (Foucault, 1980, pp. 81–83).

On the one hand, the chapter uncovers historical contents buried in the governing approach of finding "technical solutions" to the perceived "refugee problem": the persistent coloniality behind the dogmatic 1951 Refugee Convention. On the other hand, it challenges this approach by reactivating disqualified knowledges, which may open new (legal) subjectivities.

As such, this genealogical analysis proceeds along two axes: historical knowledge of struggles and the insurrection of knowledges.

As for the first axis, the chapter traces the history of struggles evolved out of empirical controversies, terminological disputes, and struggles of actors to conquer this new policymaking area.

The analysis of historical context is conducted through document-based research among the following secondary sources: research publications, research projects, official reports by international organizations and major agencies/institutions committed to the refugees/migrants, documents of proposals/ongoing initiatives to fill the legal gap, and documents/reports produced by non-governmental organizations. The references were located by systematically tracing sources cited in these documents found in libraries and internet searches. The choice of focusing on secondary sources to look at existing definitions and terms used is justified by the absence of a legal definition of "climate refugees." The main findings of this genealogical examination are represented in Table 1.1 with no pretense of offering a fully comprehensive classification. It lists 15 labels referenced from among the many more consulted. The terms have been selected as relevant according to the principle of *Entstehung* (emergence), described by Foucault as "the principle and the singular law of an apparition" (Foucault, 1977, p. 148), i.e., the first time the different terms appeared. The primary aim is to show the emergence of the term, conceived as the moment of arising, its evolution over time, and the different scenes determining the nomenclature shift.

Table 1.1 A general survey of the dispute over terms and definitions in the "Climate Refugees" discourses.

Nomenclature	Author/Institution	Definition	Year
Ecological displaced persons	William Vogt	"The people – scores of millions of people – who are using the land in disregard of its capabilities are Displaced Persons in a much more serious sense than the few hundred thousand in European refugee camps. They are displaced in the ecological sense. They can feed and clothe themselves, and supply food, fibers, charcoal, and wood to cities only by destroying the land on which they live and resources associated with it. . . . Scores of millions of them must be moved – down the eroding slopes, out of the degenerating forests, off the overgrazed ranges – if they are not to drag ever lower the living standards of their respective countries – and the world. The solution of the problem of European DPs is simple in comparison with that of the ecological DPs" (Vogt, 1949, p. 107).	1948
Ecological Refugees	Lester Brown/World Watch Institute	"As human and livestock populations retreat before the expanding desert, these ecological refugees create even greater pressure on new fringe areas, exacerbate the processes of land degradation, and trigger a self-reinforcing negative cycle of overcrowding and overgrazing in successive areas. When the inevitable drought sets in, as one did in the early seventies, this deteriorating situation is brought to a disastrous climax for the humans who perish by the hundreds of thousands, for livestock, which die in even greater numbers, and for productive land, which is destroyed" (Brown et al., 1976, p. 39).	1976
Economic Refugees	Kathleen Newland	"Throughout history people have been driven from their homes by wars or ecological catastrophes. . . . The voluntarism of these migrants' moves may be qualified by desperation and lack of alternatives, yet the force that expels them is usually not the force of arms but rather the force of circumstance. They are, in a sense, economic refugees" (Newland, 1979, p. 5).	1981
Environmental Refugees	Essam El-Hinnawi	Those people "who have been forced to leave their traditional habitat, temporarily or permanently, because of a marked environmental disruption (natural and/or triggered by people) that jeopardized their existence and/or seriously affected the quality of their life" (El-Hinnawi, 1985, p. 4).	1985
Environmental migrant	Astri Suhrke and Annamaria Visentin	Environmental migrant is a person who "makes a voluntary, rational decision to leave a region as the situation gradually worsens there. In that decision, environmental deterioration may be only one factor among others" (Suhrke & Visentin, 1991, p. 73).	1991

(Continued)

Table 1.1 (Continued)

Nomenclature	Author/Institution	Definition	Year
Environmentally displaced persons	Reinhard Lohrmann United Nations High Commissioner For Refugees (UNHCR), International Organization for Migration (IOM), Refugee Policy group (RPG)	Environmentally displaced persons are "those who can no longer gain a livelihood in their homelands because of soil erosion, deforestation, desertification, drought, chemical contamination or other related collapses in natural carrying capacity, whether short-term or long-term. Not all environmentally displaced persons flee their countries (many remain internally displaced), but a key feature is that they move because they have no other choice" (Lohrmann, 1996, pp. 335–336).	1996
Ecomigrants	William B. Wood	"Ecomigrants, unlike 'environmental refugees,' are not necessarily violently displaced. They are closely entwined with expectation-raising and frequently disruptive process of economic development. . . . Ecomigrants would include those who move voluntarily to new areas to exploit natural resources. All too often, however, these same ecomigrants are forced to leave when the resources they depend on are destroyed or severely degraded" (Wood, 2001, pp. 46–47)	2001
Environmental migrant	IOM	"Environmental migrants are persons or groups of persons who, for compelling reasons of sudden or progressive changes in the environment that adversely affect their lives or living conditions, are obliged to leave their habitual homes, or choose to do so, either temporarily or permanently, and who move either within their country or abroad" (IOM, 2007, para. 6).	2007
Climate change refugees	Bonnie Docherty and Tyler Giannini	"Climate change refugee as an individual who is forced to flee his or her home and to relocate temporarily or permanently across a national boundary as the result of sudden or gradual environmental disruption that is consistent with climate change and to which humans more likely than not contributed" (Docherty & Giannini, 2009, p. 361)	2009
Climate Refugees	Biermann and Boas	"Climate refugees" as "people who have to leave their habitats, immediately or in the near future, because of sudden or gradual alterations in their natural environment related to at least one of three impacts of climate change: sea-level rise, extreme weather events, and drought and water scarcity" (Biermann & Boas, 2010, p. 67).	2010

Term	Source	Year	Definition
Environmentally displaced persons	International Organization for Migration (IOM)	2011; 2014	Persons who are displaced within their own country of habitual residence or who have crossed an international border, and for whom environmental degradation, deterioration or destruction is a major cause of their displacement, although not necessarily the sole one (IOM, 2011, p. 34, 2014, p. 13).
Climate displaced persons	Peninsula Principles	2013	"Climate displaced persons" means individuals, households or communities who face or experience climate displacement (Peninsula Principles, Principle 2(c)).
Environmentally-displaced persons	Draft Convention on the International Status of Environmentally-Displaced Persons (University of Limoges)	2013	Environmentally displaced persons are defined by Article 2 of the Draft Convention as "individuals, families, groups and populations confronted with a sudden or gradual environmental disaster that inexorably impacts their living conditions, resulting in their forced displacement, at the outset or throughout, from their habitual residence."
Climate-change migrants	Andrew Baldwin	2013	"A human being who, either forcibly or voluntarily, migrates either temporarily or permanently from their home as an immediate or indirect result of climate change or the possibility of climate change" (Baldwin, 2013, p. 1475)
Disaster-displaced persons	Nansen Initiative	2015	The term "disaster displacement" refers to situations where people are forced or obliged to leave their homes or places of habitual residence as a result of a disaster or in order to avoid the impact of an immediate and foreseeable natural hazard. Such displacement results from the fact that affected persons are (i) exposed to (ii) a natural hazard in a situation where (iii) they are too vulnerable and lack the resilience to withstand the impacts of that hazard. It is the effects of natural hazards, including the adverse impacts of climate change, that may overwhelm the resilience or adaptive capacity of an affected community or society, thus leading to a disaster that potentially results in displacement (The Nansen Initiative, 2015, Para 16; see also https://disasterdisplacement.org/the-platform/key-definitions).
Climate displacement/Displacee(s)	Simonelli	2016	Displacees are "those who will be forced to leave their current homes due to the continual environmental deterioration and secondary concerns (those affecting their livelihoods or having other economic and social impacts) from the processes of climate change, migrating inside or outside of their home country" (Simonelli, 2016, p. 53).

(Continued)

Table 1.1 (Continued)

Nomenclature	Author/Institution	Definition	Year
Survival migrants	Betts	Survival migrants are "persons who are outside their country of origin because of an existential threat for which they have no access to a domestic remedy or resolution" (Betts, 2016, p. 23).	2016
Climate migrants	Byravan and Rajan	Climate migrants are those who are displaced because of the effects of climate change, seen, for example, in parts of Africa and Asia when there is a drought or severe flooding (Byravan & Rajan, 2017, p.109)	2017
Climate exiles	Byravan and Rajan	Climate exiles are a special class of climate migrants who will have lost their ability to remain well-functioning members of political societies in their countries, often through no fault of their own. Examples include people from the Pacific islands, many of whom have been forced to evacuate their island nations, as their lives are no longer viable due to rising seas. Furthermore, while most climate migrants will be internally displaced people, or have the opportunity of returning, climate exiles will be exiled from their nation state and will not be able to return since their nations may no longer be viable (Byravan & Rajan, 2017, p. 109)	2017
Displacees	Baldwin, Fröhlich, and Rothe	Those displaced by climate change are defined as "displaccees of a globalized network of intersecting mobility regimes fuelled by fossil fuel extraction" (Baldwin et al., 2019, p. 291).	2019

Source: Descriptive

The second axis concerning the insurrection of knowledges constitutes the *pars construens* of the chapter. It consists of confronting people involved in different climate disruptions with legal categories to identify the legal gap and show how the silence of law is a vulnerability factor in itself.

Ultimately, the concept of subjugated knowledges is key to constructing the otherness of climate refugees within a pragmatical approach to existence built upon the reactivation of local knowledges by the practices of non-state actors.

Historical knowledge of struggles

The heterogeneity and multi-causality of climate-induced migration: empirical controversies

Who to include in the calculation remains a major obstacle for scholars. The international refugee definition is unlikely to apply to people fleeing environmental disruptions as they do not meet its main requirements: alienage (most of them move internally) and persecution (the limited set of persecution grounds does not include "environmental" persecution).[2] Nor have definitions proposed by scholars been adopted as common working definitions. In the absence of such a common working definition, assuming a broader or narrower criterion can change results significantly. Further, the lack of an agreed methodology, reliable data (especially in developing countries), and consistent statistical systems render the issue mostly speculative, with empirical studies more likely to offer "guesstimates" than rigorous estimates (Baldwin, Methmann, & Rothe, 2014; Gemenne, 2011a). In light of this, it should come as no surprise that critical questions surrounding current numbers and future estimates of "climate refugees' remain largely unaddressed, having received controversial answers so far.

The first attempt of calculation was advanced by the 1985 United National Environmental Programme (UNEP) report by El-Hinnawi, which suggested the figure of 30 million displaced people by environmental change worldwide. Three years later, a key text providing estimates of the then-current numbers and future flows of "environmental refugees" was the Worldwatch Institute report by Jodi Jacobson (1988). In that report, she stated that there were 10 million "environmental refugees" at that time, and provided the first prediction that a one-meter rise in sea level could produce up to 50 million "environmental refugees" (Jacobson, 1988). In particular, she mentioned the case of Bangladesh: "eventually, the combination of rising seas, harsher storms, and degradation of the Bengal delta may wreak so much damage that Bangladesh, as it is known today, may virtually cease to exist" (Jacobson, 1989, p. 89). Remarkable warnings also came from the first Assessment IPCC 1990 report, noting that human migration might be one of the gravest effects of climate change (IPCC, 1990, pp. 55, 103).

Norman Myers (Myers, 2005; Myers & Kent, 1995) calculated that at least 25 million "environmental refugees" already existed in 1995 and predicted an increase of up to 200 million by 2050. However, there were methodological pitfalls and his calculations have been criticized. Gemenne, amongst others, noted that Myers' did not provide an accurate estimate. It described a stock (the total number of people displaced in a given place at a specific moment) rather than a flow (the number of new displacements or reported returns over a period of time) and people returned or resettled elsewhere were not included at all. Nor did it distinguish between different types of environmental changes as drivers of migration, failing to represent the complexity of human mobility, focusing on displacement (Gemenne, 2011a). These pitfalls have not prevented scholars from quoting his estimate;[3] the influential 2006 Stern Review cited Myers' predictions, albeit highlighting his estimate had not been rigorously tested.

All the predictions in circulation[4] reveal a common disagreement about the scale of future migration. The disagreement stems not only from the lack of a common working definition but also from the different views on the link between climate change and migration (Baldwin, 2013). As first observed by Suhrke (Suhrke, 1994), the literature polarized the debate in two opposing perspectives from the start: the maximalist (usually supported by natural scientists), and the minimalist (mostly argued by social scientists) views (Bettini, 2017; Jakobeit & Methmann, 2012; Klepp, 2017).[5] Whereas maximalists (El-Hinnawi, 1985; Jacobson, 1988; Myers & Kent, 1995) hold that environmental change is a direct cause for large-scale displacement of populations, minimalists (Black, 1998; Castles, 2002; Kibreab, 1997; Suhrke, 1994; Wood, 2001) affirm that there are some reasons for caution. In particular, the latter argue that climate change can contribute to migration, but analytical fallacies and empirical shortcomings make this link less obvious. In their view, the social, economic, and political contexts of migration should be given more consideration. The divide at stake is the importance of environment as *sole factor* (driver) or *one among different drivers* for migration.[6] Today, this dispute seems to have leaned in favor of minimalists as the 2011 Foresight Report provided sufficient evidence of the multi-causal nature of climate-induced migration (Baldwin et al., 2014). As such, recent studies have agreed upon moving from mono-causal environmental "push" theories to greater integration of context and multi-level analysis (Piguet, 2019).

However, disagreement remains on the magnitude of the phenomenon and the methods used to provide evidence (Gemenne, 2011b; McLeman & Gemenne, 2018). The reasons for such disagreement can generally be ascribable to the following aspects.

First of all, the absence of a clear definition of what constitutes a "climate refugee" creates uncertainty regarding who should be included in calculations: a broader or narrower definition plays a significant role in generating larger (or smaller) numbers. This subjective criterion defining "climate refugee" creates variables, such as types of migration, creating further obstacles. One such variable is voluntary/forced, the most emblematic case is that of the so-called *trapped populations*. Described in the 2011 Foresight report, *trapped populations* are those who cannot migrate and are unable to leave (Foresight, 2011). More recently, Zickgraf explored immobility as both an inability to leave and a willingness to stay despite environmental change (Zickgraf, 2019).

In addition to the problem of defining criteria for "a climate refugee" another methodological issue is that data sets regarding migration tend to be incomplete, a problem exacerbated in developing countries. In those contexts, the lack of data or reliable data sources poses a major obstacle for accurate analysis (Baldwin, 2013). Also, methodological issues arise from the so-called ecological fallacy versus atomistic fallacy dispute. In Piguet's words, "ecological fallacy" means that "correlations measured at the aggregated level might not hold true at the individual level," while "atomistic fallacy" shows that analyses strictly centered on the individual runs the risk of missing the context in which behavior takes place (Piguet, 2010, pp. 518–519). Global figures for environmental migration provided so far have been criticized for their deterministic bias. Above all, they fail to take into due account other factors significantly affecting migration behaviors (such as adaptation strategies or trapped factors) or include the timeframe for capturing long-term migration trends (Gemenne, 2011b).

In this respect, progress has been made on internal displacement by disasters and internal climate migration. Both the 2021 GRID Report by the Internal Displacement Monitoring Centre (IDMC) and the Groundswell Report by the World Bank have taken steps forward. The 2021 GRID Report offers an accurate methodology to calculate internally displaced people by disasters. By relying on different data sources and considering both stocks and flows, it produced reliable and accurate data concerning current Internally Displaced Persons by disasters. It is worth noting that, apart from the dynamic process of populations' movement captured by the IDMC data model, the report has also benefited from the adoption of an "existent," legal definition of Internally Displaced Persons established by the 1998 Guiding Principles on Internal Displacement. Instead, the 2018 Groundswell Report has the merit of introducing slow-onset climate impacts into a model of future population distribution, the first of its kind to do so. The report considers three potential climate and development scenarios by degrees of climate-friendliness. In doing so, it provides a projected number of climate migrants in Sub-Saharan Africa, South Asia, and Latin America by 2050 under the worst-case or pessimistic scenario (more than 143 million), the more inclusive development scenario (between 65 million and 105 million), and the more climate-friendly scenario (from 31 million to 72 million). In 2021, the World Bank updated these estimates, warning that slow-onset climate change impacts could force 216 million people across six world regions to move within their countries by 2050. As shown by the projections, concrete climate and development actions can significantly reduce the number of future internal climate migrants. Further examples of alternative methods seeking to overcome analytical and empirical shortcomings in the field are available in the volume edited by McLeman and Gemenne (2018), *Routledge handbook of environmental displacement and migration*. The first part of the volume explores new frontiers of environmental migration such as innovative solutions identifying potential migration patterns by scalable models, including quantitative, geospatial, and qualitative approaches to examine non-environmental factors (e.g., gender, perceptions, indigenous worldviews)

(McLeman & Gemenne, 2018). Current research on environmental migration is moving the debate beyond linear connections, introducing nuanced approaches beyond climate change and other environmental factors (Baldwin, Fröhlich, & Rothe, 2019). In doing so, cultural dimensions – like place attachment and community cohesion – contribute to expanding the notion of migration as adaptation to climate change (Adger, de Campos, & Mortreux, 2019).

To conclude, the dispute over numbers allows us to identify at least three main divides: disciplinary, methodological, and political (Gill, 2010, p. 863; de Guchteneire et al., 2011, p. 416).

The disciplinary divide mainly relies on the interdisciplinary nature of the climate-migration nexus. Given the wide range of disciplines, it is unsurprising that standpoints are divided between natural scientists studying environmental change and social scientists focusing on migration. In the past, the former have argued an alarmist perspective, portraying the impacts of climate change as the number one cause of the mass movement of people, whilst the latter have sustained the multi-causality argument on the ground that it was not possible to isolate the environmental factor as the *sole* cause to migrate. This divide has been partially overcome in favor of multi-causality and the broad disciplines involved in current research have seen this divide reduce even further. As observed by McLeman and Gemenne, natural scientists (e.g., researchers with natural hazards/socio-ecological systems/vulnerability background) draw upon social science concepts, while social scientists and refugee scholars are ever more concerned with the environmental dimensions of migration (McLeman & Gemenne, 2018). However, the question of to what extent are environmental factors working with other driving factors to influence human mobility is yet to be solved.

The methodological divide is between authors using the deductive approach to predict future displacement, and those preferring inductive research to explore the relations between climate change and patterns of vulnerability and resilience. Both methods have pitfalls resulting in the so-called ecological fallacy or atomistic fallacy; the most promising solution to this seems to be developing multi-scalar approaches integrating space, time, and context. Studies like Adger et al., which consider mobility, migration, and displacement with environmental vulnerability, have the potential to bring the debate even further forward (Adger et al., 2019).

Finally, a political divide emerges between those aiming to securitize the issue and those who are more cautious about numbers (Bettini, 2013). Whilst research such as the 2011 Foresight Report moved away from such a narrative, within migration as adaptation to changing contexts, driven by various factors including climate-related, securitization perseveres. The optimistic take over climate migration as an adjustment between inequalities (e.g., through remittances) is misleading (Bettini & Gioli, 2015; Gemenne, 2015). Far from being the cure, climate migration is a symptom of structural inequalities between developed and developing states. Furthermore, the narrative of migration as adaptation risks a social Darwinism in which the fittest, richest, whitest elites are more likely to survive compared to and at the expense of a reified, non-white other, who without agency is compelled to suffer in developing countries. Whilst current literature confirms this engulfing inequality, it does so with critical voices that are increasingly positioned around the environmental justice framework as a remedy to narratives supporting securitization, segregation, and, ultimately, environmental privilege. Whether through the subcategory of climate or mobility justice, some scholars are trying to build an alternative framework to those who see the climate migration as a problem to be solved by technical solutions (Baldwin et al., 2019; Park & Pellow, 2019; Turhan & Armiero, 2019; Whyte, Talley, & Gibson, 2019).

History of the terminological disputes

The first trace linking migration to environmental degradation can be found in Vogt's classic *Road to Survival* (1948), with its early warnings that the solution of the problem of European Displaced Persons (DPs) in the aftermath of World War II would have been simple compared with that of Ecological DPs (Vogt, 1949, p. 107). What Vogt mentions in passing comes closer to our contested concept when Lester Brown (1976) coined the term *ecological refugees* in Worldwatch Paper 5 (Brown, McGrath, & Stokes, 1976). This was soon followed by Kathleen Newland, who suggested labeling people fleeing from ecological catastrophes as *economic refugees* (Newland, 1979, p. 5). The first formal definition was introduced in the public debate by Essam El-Hinnawi in the cited 1985 UNEP report, in which El-Hinnawi defined *environmental refugees* as "those people who have been forced to leave their traditional habitat, temporarily or permanently, because of a marked environmental disruption (natural and/or triggered by people) that jeopardized their existence and/or seriously affected the quality of their life" (El-Hinnawi, 1985, p. 4). Since then, many other labels have been used, but this definition is still one of the most quoted.[7] Given the growing number of interchangeable, undefined terms used, a strand of research sought out a narrower, definitive term which restricted the number of beneficiaries receiving aid and assistance to the most (truly) vulnerable (Lassailly-Jacob & Zmolek, 1992, p. 3).

It was through this line of reasoning that the first division between *refugee* and *migrant (or other variants)* can be found. Astri Suhrke and Annamaria Visentin (Suhrke & Visentin, 1991) first criticized the overly broad definition of *environmental refugees* suggesting separating *environmental migrants* from *environmental refugees*. While the former were those who moved by choice from an area, the latter were defined as "people or social groups displaced as a result of sudden, drastic environmental change that cannot be reversed" (Suhrke & Visentin, 1991, p. 73). In a similar line of investigation, William Wood (2001) also made the distinction between *environmental refugees* and what he termed *ecomigrants*, following the same dichotomy between forced and voluntarily movement. Even in this case, however, the distinction was somewhat nebulous as the "same ecomigrants are forced to leave when the resources they depend on are destroyed or severely degraded" (Wood, 2001, pp. 46–47).

A second divide was then advanced by Docherty and Giannini, who distinguished between *environmental refugees* and *climate change refugees*

(Docherty & Giannini, 2009). In doing so, they recognized a specific room only for people in dire situations fleeing the impacts of climate change. This distinction, however, appeared equally problematic due to the challenges to prove the causal links between human activity and environmental disruptions. Over the years, this objection is becoming less relevant due to developments in attribution sciences. The 2021 IPCC report clearly states that "increases in well-mixed greenhouse gas (GHG) concentrations since around 1750 are unequivocally caused by human activities" (IPCC, 2021, p. 5). Yet, some authors raised a further issue concerning the distinction between "environmental" and "climate change refugees," arguing that there were not sufficient reasons we should treat those fleeing from environmental changes and climate change-related disruptions differently.[8]

Moving beyond such a binary logic, other authors introduced a tripartite notion of "environmental refugees." In particular, Diane Bates proposed to unfold "environmental refugees" into three subcategories by the type of environmental harm (slow/rapid onset; natural/human-caused harm). In doing so, Bates distinguished between *Disaster refugees* (who flee either natural or anthropogenic disasters), *Expropriation refugees* (relocated because of an anthropogenic expropriation of their ecosystem), and *Deterioration refugees* (who leave owing to gradual anthropogenic deterioration of their environment) (Bates, 2002, pp. 469–472).[9] On the other hand, Renaud et al. suggested creating three subcategories of "environmental refugees" by the degree of the compulsion to leave: *Environmentally motivated migrants* (i.e., those who *may leave to pre-empt the worse*), *Environmentally forced migrants* (i.e., those who *have to leave to avoid the worst*), and *Environmental refugees* (those who *flee the worst*) (Renaud, Bogardi, Dun, & Warner, 2007, pp. 29–30).

Despite numerous efforts, these initial strands of research did not manage to concur and create a working definition. The second strand was equally unsuccessful. The issue of terminology further divided scholars, enriching the number of proposals and possible perspectives to define people on the move. However, unlike the first strand that focused on restricting the number of beneficiaries to ensure aid reached the most vulnerable, the second was more concerned with deleting the "climate refugee" label from academic and policymaking circles (Gemenne, 2017; Nash, 2019).

Besides this mainstream, there are a few exceptions to mention. In spite of the strong resistance of some academic and policy spheres,[10] some authors still support using the term "climate refugees" (Biermann, 2018; Gemenne, 2017; Kent & Behrman, 2018). Biermann has argued with Boas to use this term challenging the outdated UN terminology. Why should we deserve less protection to inhabitants of atolls in the Maldives "who require resettlement for reasons of a well-founded fear of being inundated by 2050" (Biermann & Boas, 2008, p. 13) than those who flee political persecution? Biermann has reaffirmed this position in a more recent contribution in Behrman and Kent's volume *'Climate Refugees': Beyond the Legal Impasse?* (2018) in which he agrees upon the need to move beyond the monopolization of the 1951 Refugee Convention over the existence of novel types of refugees (Biermann, 2018). Similarly, Behrman and Kent's arguments are based on a critical historical perspective. In their view, the past offers many examples of evolution and extension of the 1951 Refugee Convention (see also Chapter 2 of this volume). The 1967 Protocol removed geographical and temporal limitations,[11] opening up new frontiers out of a Euro-centric post-WWII perspective. Expanded definitions of a refugee have been provided by the OAU Convention (Africa, 1969) and the Cartagena Declaration (Latin America, 1984). Both definitions included *events or circumstances seriously disturbing public order*, thus broadening the definition to those compelled to flee because of war, violence, and massive violation of human rights. Nor is the present less equipped, having created *ad hoc* typologies of refugees such as "war refugees," protected by the 1949 Geneva Convention Relative to the Protection of Civilian Persons in Time of War (IV), and "Palestine refugees," currently under the mandate of the United Nations Relief and Works Agency (Kent & Behrman, 2018). If all this is the case, why would we consider the 1951 Refugee Convention an insurmountable obstacle? Why should we believe "that there is a single, immutable legal category of the refugee in international law" (Behrman & Kent, 2018, p. 11)? Why should we deny such status and legal protection to the many who might want to have the right to be defined as refugees (Kent & Behrman, 2018)? Gemenne has provided further reflections on the significance of using the term "climate refugees." The author holds that in the context of the Anthropocene, "climate change is a form of political persecution" against the most vulnerable (Gemenne, 2017, p. 396). In his view, the term "climate refugees' has to be used outside legal understandings to re-politicize the debate and offer meaningful protection to the most vulnerable. Ultimately, a refugee is someone that seeks refuge. Why should it not be the case for a person fleeing from environmental disruptions? To ensure better protection, the term refugee becomes preferable to migrant. "Migrant' has become a negative label, due to the current populist, xenophobic narratives polarizing "refugees" as victims deserving international protection rather than opportunistic "migrants" who remain under their home State's protection.

This polarization has encountered different reactions among authors rejecting the term "climate refugees." According to Mayer, it does not justify Gemenne's conclusion (Mayer, 2018). Rather, the fact that migrants' rights are overlooked in most cases should lead to a call for more systematic protection of the human rights of all migrants in the first place (Mayer, 2016, 2018). Mayer believes that new terms are not needed for this phenomenon and puts into question the very existence of *climate* migrants/refugees as a distinct category in itself.

As he points out,

> 'environmental refugees' are not unfortunate migrants omitted by an other-
> wise fair and protective world order, but they are part of large categories of

vulnerable migrants, induced by a cluster of causes, who are generally not protected under international or national law.

(Mayer, 2012, p. 15)

Above all, climate change is not the direct cause of migration and does not produce any distinct population of *climate* migrants/refugees (Mayer, 2016, 2019). Conversely, it can only exacerbate heterogeneous forms of migration, thus playing more as an aggravating circumstance than the direct cause of a vast array of migration scenarios.

From a different perspective, other scholars argue the need to move beyond legal categories of migrants and refugees[12] to better represent the complexity of current human mobility (Bank, Fröhlich, & Schneiker, 2017; Fröhlich, 2017). In the special issue *Critical Perspectives on Human Mobility in Times of Crisis* published on *Global Policy Journal* (2017), Fröhlich has extensively considered the importance of having a historical perspective when dealing with today's human mobility. In times of ecological crisis, current mobility is still shaped by colonial power relations. The rejection of an "other" seeking refuge uncovers the violence behind Western identity creation based on the differentiation of citizens from non-citizens. This differentiation emerges from national discourses portraying the "other" as "victim" (humanitarian frame) or "threat" (security narrative). Both representations of "climate refugees," indeed, reaffirm the need to anchor Western identity on the (colonial) idea of its superiority over non-Western cultures (Fröhlich, 2017). Thus, colonial borders are not only geographical and physical but also virtual, discursive borders. On top of that, the Global North is still benefiting from GHG emissions while accumulating/exploiting resources and commodities from the Global South. Thus, further coloniality emerges from climate-induced migration from the Global South, where the impacts of climate change and extractivism are already visible and disruptive.

At the current stage of knowledge, the complexity of today's migration is not represented in the codified categories of "refugee" and "migrant," so that the figure of "climate refugees" is still perceived as not fully understandable, and so as a potentially dangerous "other" (Fröhlich, 2017).

In the issue *From climate migration to Anthropocene mobilities: shifting the debate*, which appeared in the Journal *Mobilities* (2019), Fröhlich, Baldwin, and Rothe propose a way forward. They reject the term "climate refugees" and suggest re-characterizing those displaced by climate change as "displacees of a globalised network of intersecting mobility regimes fuelled by fossil fuel extraction" (Baldwin et al., 2019, p. 291). In doing so, they replace the taken-for-grant assumption of mobility as a function of climate change, with a mobility justice perspective. In this new frame, capitalism and its fossil-fueled infrastructures are re-positioned at the center of concern about climate change and displacement (Turhan & Armiero, 2019).

Similarly, the same term, "displacees," has also been used by Simonelli (2016). In her analysis, centrality is given to the notion of *displacement* conceived as a passive process triggered by an event that moves someone involuntarily. On this basis, Simonelli advances a working definition of the terms "climate displace-ment" or 'displacee(s)" for describing

> those who will be forced to leave their current homes due to the continual environmental deterioration and secondary concerns (those affecting their livelihoods or having other economic and social impacts) from the processes of climate change, migrating inside or outside of their home country.
>
> (Simonelli, 2016, p. 53)

Another attempt to move beyond legal categorization has been made by Bettini, who brings the climate refugees/migrants dilemma into the psychological realm. Whether as "the return of the Other" who threatens the white, privileged Western elites (mainstream), "the return of the oppressed" seen as both vulnerable and poten-tial, political agent (red-green), or a "symptom of the Anthropocene" (post-human), the figure of climate refugee/migrant can be framed through the lens of Lacan's twofold conception of the symptom. Having singled out three sets of discourses (the mainstream, red-green, and post-human), the author concludes that what they all have in common is "the possibility to 'work' on the symptom, to bring back the repressed onto the scene and to move on from there" (Bettini, 2019, p. 341).

To sum up, recent debates have added further perspectives and reasons for or against the use of the term "climate refugees" compared to the criticisms raised in the past. What seems the dominant position is that the term "climate refugees" is, ultimately, a legal misnomer and has to be rejected. As a result, the 2011 Foresight report's line seems the most influential today. Even the major agencies/institutions committed to refugees/migrants firmly reject the term "climate refugees" as not having any basis in international refugee law.[13]

In compliance with the New York Declaration for Refugees and Migrants, unanimously adopted by all 193 Member States of the United Nations, the 2018 Global Compact for Safe, Orderly and Regular Migration is made up of two separate instruments: Global Compact on Refugees and the Global Compact on Migration. From the outset, its preamble recognizes that migrants and refugees are distinct groups governed by separate legal frameworks (see Chapter 2). In doing so, it clarifies that only refugees are entitled to the specific international protection defined by international refugee law (the 1951 Refugee Convention).

In this legal architecture, people fleeing due to climate change impacts are not included in either of those legal instruments. Indeed, climate change, natural dis-asters, and environmental degradation are merely mentioned amongst the adverse drivers and structural factors that compel people to leave their country of origin. They do, however, not constitute the basis to provide room for a different catego-rization. In other words, the 2018 Global Compact for Safe, Orderly and Regular Migration has reaffirmed:

1   the migrant/refugee dichotomy;
2   the absence of a distinct category of *climate* migrant/refugee.

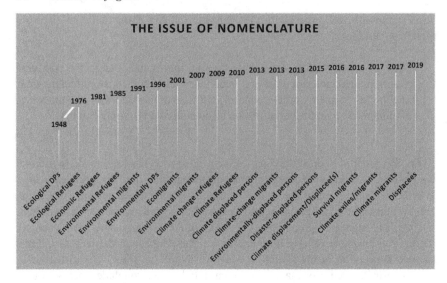

Figure 1.1 The issue of nomenclature.

Source: Descriptive

Competing for the future: the struggles of policymakers and the shift of nomenclature

A further line of analysis concerning the terms adopted for the issue of "climate refugees" relies on the shift of the nomenclature.[14]

Figure 1.1 illustrates how easily the term *refugee* alternates with *migrant or other variants* which have, in recent years, developed, varied, and complemented each other. While terms are more likely to fluctuate over the years from one group to another, one can, however, observe a slight abandonment of the term *refugee*. After 2010, *migrant* and *displaced* related expressions seem the most recurrent.

To explain this shift in the nomenclature, a watershed could be considered the 2009 UNHCR report, affirming that "UNHCR has serious reservations with respect to the terminology and notion of environmental refugees or climate refugees. These terms have no basis in international refugee law" (UNHCR, 2009, p. 8).[15] Further, the peak of the internal displacement of persons due to natural disasters (from 2008 to 2016) was achieved in 2010 with 42.4 million displaced persons. This data could explain humanitarian challenges posed on the three major institutions/organizations committed to refugees/migrants and the urgency to find an operational term to take action.[16] 2010 was also the year of the COP 16 (Cancun) Decision 1/CP.16. The relevance of the Cancun Agreement relied, first of all, on its paragraph 14 (f) being the first agreed-upon mention of migration (Nash, 2018). As established in paragraph 14 (f), all Parties were invited to undertake actions on adaptation including "[m]easures to enhance understanding, coordination and cooperation with regard to climate change induced displacement, migration and planned relocation, where appropriate, at the national, regional and international levels" (UNFCCC, 2011). Second, it recognized for the first time UNFCCC as the home for policy making on climate migration instead of the other potential policy making player, i.e., UNHCR and refugee regime (Kent & Behrman, 2018). Finally, the Cancun Adaptation Framework also had the merit to call for enhanced adaptation actions to reduce vulnerability and build resilience in developing country Parties.

In light of this, the shift from the term "climate refugees" to "climate migrants" is not surprising. The term "climate refugees" was more in line with/supported by security narratives focused on alarmist predictions of floods of "climate refugees" knocking on Western doors (Bettini, 2013). As highlighted by Baldwin et al., since the early 2000s, the UN Security Council, the UN General Assembly, NGOs, and think tanks operating in the security field have emphasized the security implications of climate-induced migration (Baldwin et al., 2014). Following the Cancun Agreement (indicating UNFCCC as policy making player), the 2011 Foresight Report's publication provided evidence that climate migration was a multi-causal phenomenon, thus dismantling security narratives providing alarming numbers based on mono-causality. The debate then moved from refugee politics and security domains toward a more nuanced understanding of human mobility, including mobility responses to the impacts of climate change: migration, displacement, planned relocation, and immobility. Thus, the term "climate refugees" was slightly replaced with "climate migrants" supported by the "new" migration as adaptation narrative rooted in the notion of resilience (Bettini, 2014, 2017).

According to Bettini, despite of the apparent diversity, both narratives appear substantially complementary to neoliberal governance embedded in a biopolitical framework. Those two narratives are faces of the same coin, as the notion of resilience uncovers a "new lexicon of security." In his view, only resilient heroes who successfully adapt to Western societies are welcome, while those who fail to adapt are still conceived as potential threats to developed countries of destination. Behind the "mild tones" of resilience build upon the migration-development nexus, this narrative produces a triage logic, still excluding the unfit under the name of biopolitics (Bettini, 2014).

Despite the criticism raised against the optimistic neoliberal takeover of the development and migration debate (Bettini & Gioli, 2015; Gemenne, 2015), the mainstream successfully portrayed migration as a useful tool to foster development. In this context, it is not surprising that IOM began to emerge as a key actor. Climate migration has been IOM's agenda since the early 1990s when IOM's Member States expressed interest in understanding climate migration's impacts on countries of origin, transit, and destination. Since then, IOM has worked extensively on terminology and institutional views to respond to the most vulnerable Member States' needs (Chazalnoel & Ionesco, 2018). The IOM 2007 working definition of climate migration represented a milestone which has been widely applied in this field of research. After the 2010 Cancun Agreement, with the link of climate change and migration became a central topic in IOM's strategic view and gained momentum in 2015 with the creation of Migration, Environment and Climate Change division (MECC) further acknowledging IOM as a key actor in the climate migration field (Nash, 2018).

A further shift in nomenclature can be observed from 2013 when the term "climate displaced," and related variants started to emerge. This further variation can be explained by some events: the rise of the Nansen Initiative in 2012, the establishment of Warsaw International Mechanism in 2013, the creation of a UNFCCC-led Task Force on displacement in 2015, and the adoption of the Sendai Framework for Disaster Risk Reduction in 2015. The Nansen Initiative is a state-led consultative process launched by Switzerland and Norway in 2012 with the aim of protecting people displaced across borders in the aftermath of disasters. This bottom-up process culminated in 2015 with the adoption of the protection agenda. It is well-situated in the international and regional recognition of the challenges related to human mobility in the context of disasters and climate change introduced by the Sendai Framework for Disaster Risk Reduction (UNISDR, 2015). The recommendations included in the protection agenda provided the impulse to establish the Platform on Disaster Displacement (PDD) in 2016, aimed at supporting states during implementation. In parallel, the 2015 Paris climate negotiations requested the Warsaw International Mechanism (WIM), established in 2013 to work on loss and damage having migration in its initial work plan, to set a Task Force on displacement. Its main task was to recommend integrated approaches to avert, minimize, and address displacement related to climate change's adverse impacts.

The migration phase is therefore gradually shifting toward a major engagement with the topic of displacement both with the acknowledgment of the failure of refugee/migrant terms to represent the human mobility's complexity (Fröhlich, 2017) and also, the availability of more accurate methods of calculation allowing reliable data collection, at least with internally displaced people by natural disasters (e.g., progress made by the Internal Displacement Monitoring Centre, IDMC). Evidence of this terminological shift can be found in the terms adopted by recent initiatives: Disaster-displaced persons (The Nansen Initiative, 2015), Climate displaced persons (Displacement Solutions, 2013), and Environmentally-displaced persons (Prieur et al., 2008). The three initiatives mentioned earlier differ in scope. While the Nansen Initiative applies to displaced persons who cross the border and Peninsula Principles (the first formal policy of its kind in the world) to internally displaced persons, the 2013 Draft Convention proposed by the University of Limoges is the only instrument applicable to both categories (see Chapter 3 of this volume).

The move toward displacement certainly has some advantages, but there are reasons for caution. The term offers a way out of the difficulties of distinguishing between voluntary or forced movement, collecting reliable data, and setting apart anthropogenic climate change impact, particularly relevant as the context of disasters and climate change encompass both anthropogenic and non-anthropogenic climate change impacts. Some natural disasters are sudden and violent, leaving little doubts on the involuntary and unforeseen movement of the people affected, making displacement a useful term. However, the concept of displacement may lead to a generalized *climatisation*[17] of the issue (Maertens & Baillat, 2018), thus obscuring the underlying responsibility issues and its political roots. Further, the domain of displacement is likely to conceive the affected populations as passive agents disregarding their reaction to this phenomenon.

A critical perspective to this conceptualization has been recently advanced by the issue mentioned earlier: *From climate migration to Anthropocene mobilities: shifting the debate* (2019). In this issue, some authors questioned and partially addressed the need to re-position the debate around the geohistorical conditions of mobility (capitalism along with its fossil-fueled infrastructures) (Bettini, 2019; Turhan & Armiero, 2019). The concept of mobility justice is mobilized to provide an expanded reading of the link between climate change and human mobility. Above all, the issue aims to challenge institutionalized relationships between these two phenomena provided within the United Nations Framework Convention on Climate Change, specifically the Warsaw International Mechanism on Loss and Damage. In particular, it opposes the speculation that "climate change will proliferate various forms of human mobility, specifically migration, displacement and resettlement, in ways that threaten international security, yield new humanitarian crises, and tax an already overburdened refugee regime" (Baldwin et al., 2019, p. 290).

Although terms and narratives have changed over the years, the narrative of securitization continues to re-emerge. The violence of Western identity based on the differentiation of citizens from non-citizens (Fröhlich, 2017) materializes in Pellow's understanding of the current environmental privilege of the few, who freely circulate and want to secure *privatized places untouched by global turmoil*, opposed to the many, who suffer in place or in other developing countries. In Pellow's words,

> the idea of mobility justice is intended to reveal and challenge the ways in which social inequalities shape, restrict, and criminalize mobilities for certain bodies while normalizing and enabling mobilities for others through discursive and material systems of racial, gender, sexual, and national differences and border making.
>
> (Park & Pellow, 2019, p. 396)

Thus, the border making realized by restricting immigrants' movement across national borders allows securing environmental amenities to the white and wealthy elites at the expense of the environments and climates of the others, non-white and poor (see also Chapter 4). Yet, this form of environmental privilege can be well observed in "pan-European efforts to seal off the EU's external borders against immigration from non-EU states, while at the same time normalizing the mobility of European migrants" (Fröhlich, 2017, p. 7). This is reflected in how borders in almost all the Global North have militarized security. Referring to this militarization, Parenti coined the term "armed lifeboat" for rich countries using their privilege to protect their own elites, while shutting out "climate refugees" (Parenti, 2011). The forthcoming struggle for life with the fittest white, rich elites surviving at the expense of the unfit non-white poor may find a way out by the new insights offered by Survival Migration (Betts, 2016).

Survival migration, as a new perspective, shifted the focus from the substantially undefined persecution of the 1951 Refugee Convention, toward access to fundamental rights necessary for survival. According to Betts, survival migrants can be defined as "persons who are outside their country of origin because of an existential threat for which they have no access to a domestic remedy or resolution" (Betts, 2016, p. 23). Significantly, by this definition, existential threat is not defined by empirical environmental factors influencing mobility or the underlying causes of movement, but rather, the threat is defined by identifying and establishing a threshold of fundamental rights in the country of origin, the lack of which requires and justifies international protection.

The insurrection of knowledges – legal categories in motion

The multi-causality and complexity of climate-induced migration require multidisciplinary research bridging natural and social science more than ever. As correctly observed by Fröhlich (2017), legal categories of "refugee" or "migrant" not only are problematic but cannot fully represent the complexity of human mobility, however, this legal perspective can't be excluded or eluded. Changes in people's status as migrant/refugee do not simply correspond to the severity and type of environmental factors they experience, but exist in direct relationships with the municipalities using these terms to codify them. This leads to an inevitable question: Does the existing legal framework have the necessary tools to cope with this complexity?

The vast production of labels and definitions discussed in this chapter has demonstrated that no word can capture all this complexity alone and the fragmentation into more labels is somehow unavoidable. Overall, the terms collected so far can be clustered by three categories: refugees, migrants, and displaced persons. These are all existing categories in law with defined meaning and scope. But are these categories sufficient?

The migrant category may cover people affected by the early stages of environmental degradation caused by slow-onset events. This category also benefited from the 2018 Global Compact on Migration (GCM), framing migration in the context of the 2030 Agenda for Sustainable Development. Although it has not created a specific legal category of *climate* migrants, the GCM has reminded States their international law obligations to "respect, protect and fulfil the human rights of all migrants, regardless of their migration status.[18]" So, at least, they are entitled to human rights within the more general category of migrants.

People displaced by rapid-onset or extreme weather events can be categorized as displaced persons. Most of them do not cross the border, and in theory should be protected as Internally Displaced Persons under the 1998 Guiding Principles on Internal Displacement, the Peninsula Principles on Climate Displacement within States, and the 2009 Kampala Convention (although for the African context only). Furthermore, the recent Nansen Initiative's Agenda for Protection has protected those who cross the border in the context of climate change and disasters. However, with the sole exception of the 2009 Kampala Convention, all these legal instruments are recommendations and soft-law mechanisms lacking any legally binding force and cannot be regarded as a source for rights.

The refugee category appears the most problematic. The "futurology" of climate-induced migration implies that the case of an environmental degradation up to the form of uninhabitability was not even imagined when the 1951 Refugee Convention was designed (Kent & Behrman, 2018). Therefore, later stages of slow-onset events, including peculiar cases like sinking islands (SIDS), potentially leading to uninhabitability, do not seem to have received convincing legal answers. People experiencing these (or the fear of these) are still conceived to be forced migrants, thus having no option but coming back to their countries of origin where their lives are at risk.

However, the spectrum of uninhabitability and the consequent well-founded fear introduce a difference of kind (refugee) rather than a degree (forced migrant). While no other category but refugee may meet such a well-founded fear of suffering from life-threatening conditions in countries of origin turning uninhabitable, the current definition does not offer room for protection in its present formulation. Nor has the law convincing answers for extreme cases where the relocation is expected to be permanent up to the form of exile (Eckersley, 2015).

The UN Human Rights committee's decision of 7 January 2020, in the case of Teitiota v. New Zealand, in which Teitiota applied for refugee status on the basis of climate change, opened up a window of hope and can be well considered a legal tipping point (see Chapter 2). Although the decision ultimately rejected the claim to protection by Mr. Teitiota being not at imminent risks, the committee admitted that

> without robust national and international efforts, the effects of climate change in receiving states may expose individuals to a violation of their rights under articles 6 or 7 of the Covenant, thereby triggering the non-refoulement obligations of sending states. Furthermore, given that the risk of an entire country becoming submerged under water is such an extreme risk, the conditions of life in such a country may become incompatible with the right to life with dignity before the risk is realized.
>
> (para. 9.11)[19]

In other words, the committee nonetheless recognized that people who flee the effects of climate change and natural disasters should not be returned to their country of origin if essential human rights would be at risk on return.[20]

This historic decision justifies the attempts to overcome the legal impasse, calling for broader understanding of the term refugee. This broader interpretation is achieved by an analogical reasoning in law. It consists of two steps. The first step is to establish a similarity between the new case (i.e., "climate refugees") and the one explicitly governed by the 1951 Refugee Convention (let us call it "political refugees"). An empirical analogy is generally drawn on the basis of a *physical* quality and arrives at a probabilistic conclusion. By contrast, an argument by analogy in law is based on a *deontic* quality and arrives at a normative conclusion (Juthe, 2016). This means that one should look at the similarities or dissimilarities between climate and political refugees in the light of the *ratio legis*, which operates as the *tertium comparationis* (the third of comparison). It is the *ratio legis*, indeed, that provides the standard of comparison to establish if both cases are alike, and this leads us to the second step. Once the similarity has been established, it is employed to justify the inference of the same legal consequence. According to the principle of formal justice,[21] if two cases are relevantly similar, they ought to be treated alike. Therefore, given that the *ratio legis* underlying the international refugee law is providing international assistance to those deprived by basic rights with no recourse to home government (Shacknove, 2010), similarities between "climate" and "political" refugees can be well drawn along the line of vulnerability. Regardless of the reasons why they escape from their countries of origin (e.g., persecution, war, natural disasters), those people share the same vulnerabilities, being their essential human rights at risk upon return in their home countries (see also Chapter 4 of this volume).

While it remains to be clarified if and to what extent those displaced by climate change and disasters can find protection under the Global Compacts,[22] the silence of law concerning this issue constitutes a further factor of vulnerability. Indeed, in the lack of formal recognition under international refugee law, those people are classified as irregular migrants. Not only is this likely to exacerbate their vulnerability and basic rights deprivation, but it does not comply with the formal justice principle according to which two similar cases ought to be treated alike.

As the genealogical analysis of historical knowledge of struggles has shown, discourses on climate refugees have suffered from a form of objectification. Not only have they omitted human beings and their lives jeopardized by the impacts of climate change from the discourse on migration and climate change nexus (Nash, 2019), but they have also led the discussion to an overly narrow empirical realm more focused on demonstrating the existence of certain phenomenon with physical qualities, than having a normative reflection based on justice. Against this background, the point is not to set up a controversy about the exact nature of climate refugees in terms of truth or post-truth (Mayer, 2018), but rather in terms of justice or injustice, thus revising our understanding of a refugee given the unfair treatment reserved to those fleeing in the context of climate disruptions. As recalled by Rawls:

> Justice is the first virtue of social institutions, as truth is of systems of thought. A theory however elegant and economical must be rejected or revised if it is untrue; likewise laws and institutions no matter how efficient and well-arranged must be reformed or abolished if they are unjust.
>
> (Rawls, 1971, p. 3)

Concluding remarks: toward a decolonial environmental justice perspective

Having accomplished that (i) the main focus should be shifted from physical phenomena to people and their human rights, (ii) this issue should be dealt with by social institutions rather than systems of thought, and since (iii) it is a matter of justice rather than truth, the questions emerge: Which kind of justice are we dealing with? How and who should be doing justice for "climate refugees"?

The root causes of climate-induced migration result from the interplay of social, economic, and environmental inequalities. In particular, environmental inequalities are becoming significant drivers of climate-induced migration as those more likely to encompass the non-human world and its relationship with humans. In this view, the kind of justice I am referring to, concerning climate-induced migration, is environmental justice. In other words, by not recognizing "climate refugees" as a new legal subjectivity, the international community does not take into due account the discriminatory effect that environmental changes may have on certain groups. In doing so, it also fails to recognize differences among human groups and their worldviews, thus perpetuating a colonial, triage, state-centric approach. Above all, the history of struggles surrounding the issue of climate refugees has revealed the persistence of coloniality of power, knowledge, and being (see Chapter 4).

The coloniality of power has emerged both through (1) the racial difference between Europeans and non-Europeans aimed at facilitating the movement/migration of white-skilled rather than non-white, unfit people on the move; and (2) the use of Western institutional forms of power, such as the nation-state, in non-Western societies where different cultures and legal pluralism exist.

A further dimension of the coloniality of power evolves from the coloniality of knowledge through the power of certain policymaking actors likely to influence terminology and working definitions in use. In turn, coloniality of power and knowledge affects the individual through the coloniality of being, materialized by mechanisms of objectification aimed at jettisoning human beings and their human rights from the debate. This form of objectification of the life of the most vulnerable groups prevents the construction of a new (legal) subjectivity, thus sacrificing the other, non-white, poor (adapted from Rodriguez, 2020). In particular, this colonial, triage logic survives in the Westphalian order crystallized into the dogmatic 1951 Refugee Convention's definition. Not only is this definition one among the many others used in the past and present (Kent & Behrman, 2018), but it is embedded in an overly state-centric approach. Receiving states have the last word in accepting people seeking refuge, while sending states are considered failed states unable to ensure their citizens' basic rights. All this provides a logic of culpability where sending states are culpable for having produced refugees – i.e., problems to be solved by the international community – and people seeking refuge are culpable when they fail to integrate into host communities – they are conceived not resilient enough.

Against this background, recognizing the category of "climate refugees" would have the potential to overcome this persistent coloniality by acknowledging the existence of different worldviews and ways of being and knowing, but would not be sufficient to shift the perspective from a state-centric to a more community-sensitive approach.

Drawing on the Latin American decolonial environmental justice perspective, I hold that the search for recognition and inclusion by analogy in existing legal categories should be complemented with the construction of a new subjectivity within a pragmatical approach to existence. The construction of such a new (legal) subjectivity is to be built upon the reactivation of local knowledges through the practices of non-state actors (see Chapter 5 of this volume).

A non-state actor approach would have various advantages. In line with an environmental justice perspective, it may first introduce a community-sensitive approach more concerned with people's human rights and their relationship with the non-human world. Secondly, it may contribute to overcoming methodological and political divides. As for the methodological divide, non-state actors acting together in a coordinated process have the potential to bridge the atomistic and the ecological fallacy, being able to accommodate diversities at the individual level and implement context-tailored policies. As for the political divide, non-state actors may help overcome the persistent securitization of climate-induced migration by replacing discourses dominated by rhetorical tropes with actions able to make visible, and hopefully to remedy, the higher vulnerability of certain groups hit by climate change.

Given the lack of formal recognition of this new subjectivity, climate refugees may exist only in action, regardless of the approval of epistemic communities in the service of established regimes of thought. In the words of Foucault, "power is neither given, nor exchanged, nor recovered, but rather exercised, and . . . it only exists in action" (Foucault, 1980, p. 89).

Notes

1 As enshrined in UNHCR, *Guiding Principles on Internal Displacement*, 22 July 1998, p. 5: "internally displaced persons are persons or groups of persons who have been forced or obliged to flee or to leave their homes or places of habitual residence, in particular as a result of or in order to avoid the effects of armed conflict, situations of generalized violence, violations of human rights or natural or human-made disasters, and who have not crossed an internationally recognized State border."

2 The 1951 Refugee Convention read in conjunction with its 1967 Protocol defines a refugee as a person who: "owing to well-founded fear of being persecuted for reasons of race, religion, nationality, membership of a particular social group or political opinion, is outside the country of his nationality and is unable, or owing to such fear, is unwilling to avail himself of the protection of that country; or who, not having a nationality and being outside the country of his former habitual residence as a result of such events, is unable or, owing to such fear, is unwilling to return to it." For a critical overview of this definition, see also Chapter 2 of this volume.

3 See amongst others, Foresight Report still citing his contribution (Foresight, 2011).

4 For a detailed list of projections of future numbers of "climate refugees," see (Jakobeit & Methmann, 2012, p. 303). For a complete overview of both estimates and projections of "climate refugees," see (Gemenne, 2011).

5 Those two different views are also known and labeled as "alarmists" versus "skeptics." Cf. (Gemenne, 2011, pp. 230–239). Recently, James Morrissey has argued for a description of the literature on environmental refugees characterized in terms of "proponents" and "critics" of the "environmental refugee" concept.

6 For valuable empirical research on the environmental change-migration nexus, see the EACH-FOR project, a two-year-long research project of the European Commission (https://migration.unu.edu/research/migration-and-environment/environmental-change-and-forced-migration-scenarios-each-for-2.html#outline); cf. its findings in (de Guchteneire, Pécoud, Piguet, 2011, chap.8).

7 To date, the lively debate over the terminology and possible working definitions has generated the following set of terms: Ecological Displaced Persons, Ecological Refugees, Economic Refugees, Environmental Refugees, Environmental Migrants,

Environmentally Displaced Persons, Ecomigrants, Climate Change Refugees, Climate Refugees, Climate Displaced Persons, Climate-Change Migrants, Environmentally-Displaced Persons, Disaster-Displaced Persons, Climate Exiles, Climate Migrant.

8 Climate change mainly involves atmospheric processes, while environmental change comprises all processes that shape the environment, including changes in atmospheric temperature, geological factors (e.g., erosion, weathering, volcanism), and biological factors (e.g., invasive species). Therefore, climate change can be well considered a subcategory of environmental change.

9 Cf. (Jakobeit & Methmann, 2007, p. 11) who provided an expanded list of five categories along the dichotomy Unfreiwillig (Umweltflüchtling; Antizipierender Umweltflüchtling; Getriebene UmweltmigrantIn; Getriebene MigrantIn, u.a. wg. Umwelt) / Freiwillig (Freiwillige MigrantIn).

10 Among others, Benoit Mayer has accused those using the term "climate refugees" of being naive, post-truth supporters, and eroding the Academia's scientific authority. In his view, those authors' political engagement leads them to put the political advocacy before analytical considerations and the desired political impact over the reality's representation (Mayer, 2018, pp. 95–97).

11 Until the 1967 Protocol, the 1951 Refugee Convention was intended to protect people fleeing events occurring *before 1 January 1951* and *within Europe*.

12 On the need to move beyond the dichotomy between refugees and migrants, see also (Hamlin, 2021).

13 Leading authorities suggested avoiding the term "climate refugees" during the 2011 Nansen Conference ((NRC/IDMC), 2011). See also Dina Ionesco's – Head of the Migration, Environment and Climate Change IOM's Division – position against this term recently expressed here www.un.org/sustainabledevelopment/blog/2019/06/lets-talk-about-climate-migrants-not-climate-refugees/. Leading scholars in this field have also expressed their opposition to the use of this term (Kälin & Schrepfer, 2012; McAdam, 2009).

14 To understand "why people migrate between labels," see also (Zetter, 2007), describing the formation, transformation, and politicization of the refugee label.

15 Resistance in supporting the term "climate refugees" has already been advanced in the findings of the Symposium held in Chavannes-de-Bogis between 21 and 24 April 1996, where UNHCR, RPG, and IOM preferred using the term *Environmentally displaced persons* (Lohrmann, 1996). In 1997, UNHCR had already distinguished (true) refugees "by the fact that they lack the protection of their state and therefore look to the international community to provide them with security. Environmentally displaced people, on the other hand, can usually count upon the protection of their state, even if it is limited in its capacity to provide them with emergency relief or longer-term reconstruction assistance." (UNHCR, 1997: Chap.1, Box 1.2).

16 For a critical overview of the policies implemented to deal with disaster management, see (Gemenne, 2011, pp. 243–248).

17 The term *climatisation* can be defined as the growing trend for presenting formerly unrelated issues to the climate problem through a "climatic lens." Such a trend usually aims to conceal the social, economic, and political roots of those issues.

18 Global Compact for Safe, Orderly and Regular Migration, para 11.

19 *Ioane Teitiota v. New Zealand (advance unedited version)*, CCPR/C/127/D/2728/2016, UN Human Rights Committee (HRC), 7 January 2020, available at: www.refworld.org/cases,HRC,5e26f7134.html [accessed 17 July 2020]

20 For a critical perspective on this decision, see www.qil-qdi.org/the-teitiota-case-and-the-limitations-of-the-human-rights-framework/ [accessed 10 September 2021].

21 The principle of formal justice can be summarized as follows: "treat relevantly similar cases alike and relevantly unlike cases unlike" (Reidhav, 2007, p. 48, as cited in Juthe, 2016, p. 70).

22 Cf. GCR para 12, GCM para 18 (h), GMG p. 6.

Reference list

Adger, W. N., de Campos, R. S., & Mortreux, C. (2019). Mobility, displacement and migration, and their interactions with vulnerability and adaptation to environmental risks. In F. Gemenne & R. A. McLeman (Eds.), *Routledge handbook of environmental displacement and migration* (pp. 29–41). Routledge. https://doi.org/10.4324/9781315638843-3

Baldwin, A. (2013). Racialisation and the figure of the climate-change migrant. *Environment and Planning A, 45*(6), 1474–1490. https://doi.org/10.1068/a45388

Baldwin, A., Fröhlich, C., & Rothe, D. (2019). From climate migration to anthropocene mobilities: Shifting the debate. *Mobilities, 14*(3), 289–297. https://doi.org/10.1080/17450101.2019.1620510

Baldwin, A., Methmann, C., & Rothe, D. (2014). Securitizing 'climate refugees': The futurology of climate-induced migration. *Critical Studies on Security, 2*(2), 121–130. https://doi.org/10.1080/21624887.2014.943570

Bank, A., Fröhlich, C., & Schneiker, A. (2017). The political dynamics of human mobility: Migration out of, as and into violence. *Global Policy, 8*(Suppl. 1), 12–18. https://doi.org/10.1111/1758-5899.12384

Bates, D. C. (2002). Environmental refugees? Classifying human migrations caused by environmental change. *Population and Environment, 23*(5), 465–477. https://doi.org/10.1023/A:1015186001919

Behrman, S., & Kent, A. (2018). *Climate refugees: Beyond the legal impasse?* Abingdon: Routledge. https://doi.org/10.4324/9781315109619

Bettini, G. (2013). Climate barbarians at the gate? A critique of apocalyptic narratives on "climate refugees." *Geoforum, 45*, 63–72. https://doi.org/10.1016/j.geoforum.2012.09.009

Bettini, G. (2014). Climate migration as an adaption strategy: De-securitizing climate-induced migration or making the unruly governable? *Critical Studies on Security, 2*(2), 180–195. https://doi.org/10.1080/21624887.2014.909225

Bettini, G. (2017). Where next? Climate change, migration, and the (Bio)politics of adaptation. *Global Policy, 8*, 33–39. https://doi.org/10.1111/1758-5899.12404

Bettini, G. (2019). And yet it moves! (Climate) migration as a symptom in the anthropocene. *Mobilities, 14*(3), 336–350. https://doi.org/10.1080/17450101.2019.1612613

Bettini, G., & Gioli, G. (2015). Waltz with development: Insights on the developmentalization of climate-induced migration. *Migration and Development, 5*(2), 171–189. https://doi.org/10.1080/21632324.2015.1096143

Betts, A. (2016). *Survival migration. Survival migration.* Cornell University Press. https://doi.org/10.7591/cornell/9780801451065.003.0002

Biermann, F. (2018). Global governance to protect future climate refugees. In S. Behrman & A. Kent (Eds.), *Climate refugees: Beyond the legal impasse?* (pp. 265–277). New York: Routledge. https://doi.org/10.4324/9781315109619

Biermann, F., & Boas, I. (2008). Protecting climate refugees: The case for a global protocol. *Environment, 50*(6), 8–17. https://doi.org/10.3200/ENVT.50.6.8-17

Biermann, F., & Boas, I. (2010). Preparing for a warmer world: Towards a global governance system to protect climate refugees. *Global Environmental Politics, 10*(1), 60–88. https://doi.org/10.1162/glep.2010.10.1.60

Black, R. (1998). *Refugees, environment and development.* London: Longman.

Brown, L., McGrath, P., & Stokes, B. (1976). *Twenty-two dimensions of the population problem. Population reports. Series J: Family planning programs* (Vol. 35). https://doi.org/10.2307/1531902

Byravan, S., & Rajan, S. C. (2017). Taking lessons from refugees in Europe to prepare for climate migrants and exiles. *Environmental Justice, 10*(4), 108–111. https://doi.org/10.1089/env.2016.0026

Castles, S. (2002). *Environmental change and forced migration: Making sense of the debate* (Working Paper No. 70). New Issues in Refugee Research. Geneva: United Nations High Commissioner For Refugees (UNHCR).

Chazalnoel, M. T., & Ionesco, D. (2018). Advancing the global governance of climate migration through the United Nations framework convention on climate change and the global compact on migration: Perspectives from the international organization for migration. In Simon; Behrman & A. Kent (Eds.), *Climate refugees: Beyond the legal impasse?* (pp. 104–117). Routledge. https://doi.org/10.4324/9781315109619

Displacement Solutions. (2013, 18 August). *The Peninsula Principles on climate displacement within states.* Retrieved from https://reliefweb.int/sites/reliefweb.int/files/resources/Peninsula-Principles.pdf

Docherty, B., & Giannini, T. (2009). Confronting a rising tide: A proposal for a convention on climate change refugees. *Harvard Environmental Law Review, 33*(2), 349–403.

Eckersley, R. (2015). The common but differentiated responsibilities of states to assist and receive 'climate refugees.' *European Journal of Political Theory, 14*(4), 481–500. https://doi.org/10.1177/1474885115584830

El-Hinnawi, E. (1985). *Environmental refugees.* Nairobi: United Nations Environment Programme.

Foresight: Migration and Global Environmental Change. (2011). *Final project report.* London: The Government Office for Science. Retrieved from https://www.un.org/development/desa/pd/sites/www.un.org.development.desa.pd/files/unpd-cm10201202-11-1116-migration-and-global-environmental-change.pdf

Foucault, M. (1977). Nietzsche, genealogy, history. In D. F. Bouchard (Ed.), *Language, counter-memory, practice: Selected essays and interviews* (pp. 139–164). Cornell University Press. https://doi.org/10.2307/2905802

Foucault, M. (1980). Two lectures. In G. Colin (Ed.), *Power/knowledge: Selected interviews and other writings 1972–1977* (pp. 78–108). New York: Pantheon Books.

Fröhlich, C. (2017). A critical view on human mobility in times of crisis. *Global Policy, 8*, 5–11. https://doi.org/10.1111/1758-5899.12417

Gemenne, F. (2011a). Why the numbers don't add up: A review of estimates and predictions of people displaced by environmental changes. *Global Environmental Change, 21*(Suppl. 1), S41–S49. https://doi.org/10.1016/j.gloenvcha.2011.09.005

Gemenne, F. (2011b). How they became the human face of climate change. Research and policy interactions at the birth of the "environmental migration" concept. In P. Piguet, E. Pécoud, & A. de Guchteneire (Ed.), *UNESCO migration and climate change* (pp. 225–259). Cambridge: Cambridge University Press.

Gemenne, F. (2015, May). One good reason to speak of "climate refugees." *Forced Migration Review, 49*, 70–71.

Gemenne, F. (2017). The refugees of the anthropocene. In B. Mayer & F. Crépeau (Eds.), *Research handbook on climate change, migration and the law* (pp. 394–404). https://doi.org/10.4337/9781785366598.00025

Gill, N. (2010). "Environmental refugees": Key debates and the contributions of geographers. *Geography Compass.* https://doi.org/10.1111/j.1749-8198.2010.00336.x

Hamlin, R. (2021). *Crossing: How we label and react to people in the move.* Stanford: Stanford University Press.

IOM. (2007). *Discussion note: Migration and the environment, MC/INF/288.* Geneva: IOM.

IOM. (2011). *Glossary on Migration, International Migration Law* (Vol. n° 25, 2nd ed.). Geneva: IOM. https://doi.org/10.1017/CBO9781107415324.004

IOM. (2014). *Glossary – Migration, Environment and Climate Change: Evidence for Policy (MECLEP)*. Geneva: IOM.

Ionesco, D. (2017). *The atlas of environmental migration*. Routledge. https://doi.org/10.4324/9781315777313

IPCC. (1990). *IPCC first assessment report. IPCC first assessment report*. https://doi.org/10.1038/360292e0

IPCC. (2021). IPCC, 2021: Summary for policymakers. In B. Z. Masson Delmotte, V., P. Zhai, A. Pirani, S. L. Connors, C. Péan, S. Berger, N. Caud, Y. Chen, L. Goldfarb, M. I. Gomis, M. Huang, K. Leitzell, E. Lonnoy, J. B. R. Matthews, T. K. Maycock, T. Waterfield, & O. Yelekçi (Eds.), *Climate change 2021: The physical science basis. Contribution of working group I to the sixth assessment report of the intergovernmental panel on climate change* (pp. 1–41). Cambridge University Press.

Jacobson, J. L. (1988). Environmental refugees: A yardstick of habitability. *Worldwatch Paper, 86.*

Jacobson, J. L. (1989). Abandoning homelands. In L. Brown (Ed.), *State of the world 1989: A Worldwatch Institute report on progress toward a sustainable society*. New York: Norton.

Jakobeit, C., & Methmann, C. (2007). *Klimaflüchtlinge*. Hamburg: Greenpeace.

Jakobeit, C., & Methmann, C. (2012). 'Climate refugees' as dawning catastrophe? A critique of the dominant quest for numbers. In J. Scheffran, M. Brzoska, H. G. Brauch, P. M. Link, & J. Schilling (Eds.), *Climate change, human security and violent conflict: Challenges for societal stability* (pp. 301–314). Berlin, Heidelberg: Springer Berlin Heidelberg. https://doi.org/10.1007/978-3-642-28626-1_16

Juthe, A. (2016). Classifications of arguments by analogy part I. A comprehensive review of proposals for classifying arguments by analogy. *Cogency, 8*(2), 51–99.

Kälin, W., & Schrepfer, N. (2012). *Protecting people crossing borders in the context of climate change: Normative gaps and possible approaches*. Geneva: UNHCR.

Kent, A., & Behrman, S. (2018). *Facilitating the resettlement and rights of climate refugees: An argument for developing existing principles and practices*. Abingdon: Routledge. https://doi.org/10.4324/9781351175708

Kibreab, G. (1997). Environmental causes and impact of refugee movements: A critique of the current debate. *Disasters, 21*(1), 20–38. https://doi.org/10.1111/1467-7717.00042

Klepp, S. (2017). Climate change and migration. In *Oxford Research Encyclopedia of Climate Science*. Oxford: Oxford University Press.

Lassailly-Jacob, V., & Zmolek, M. (1992). Environmental refugees. *Refugee, 12*(1), 1–4.

Lohrmann, R. (1996). Environmentally-induced population displacements and environmental impacts from mass migrations. Conference report. *International Migration (Geneva, Switzerland), 34*(2), 335–339. https://doi.org/10.1111/j.1468-2435.1996.tb00529.x

Maertens, L., & Baillat, A. (2018). The partial climatisation of migration, security and conflict. In S. C. Aykut et al. (Eds.), *Globalising the climate* (pp. 116–134). Routledge. https://doi.org/10.4324/9781315560595-7

Mayer, B. (2012). 'Environmental refugees'? A critical perspective on the normative discourse. *SSRN Electronic Journal*. https://doi.org/10.2139/ssrn.2111825

Mayer, B. (2016). *The concept of climate migration: Advocacy and its prospects*. https://doi.org/10.4337/9781786431738

Mayer, B. (2018). Who are "climate refugees"? Academic engagement in the post-truth era. In Simon; Behrman & A. Kent (Eds.), *Climate refugees: Beyond the legal impasse?*

(pp. 89–100). Faculty of Law, Chinese University of Hong Kong: Taylor and Francis. https://doi.org/10.4324/9781315109619

Mayer, B. (2019). Definitions and concepts. In R. McLeman & F. Gemenne (Eds.), *Routledge handbook of environmental displacement and migration* (pp. 323–328). https://doi.org/10.4324/9781315638843-25

McAdam, J. (2009). From economic refugees to climate refugees? Review of international refugee law and socio-economic rights: Refuge from deprivation by Michelle Foster. *Melbourne Journal of International Law, 10*(2), 579–595.

McLeman, R., & Gemenne, F. (2018). *Routledge handbook of environmental displacement and migration.* London: Routledge.

Myers, N. (2005, May 23–27). Environmental refugees: An emergent security issue. *13th Economic Forum.*

Myers, N., & Kent, J. (1995). *Environmental exodus: An emergent crisis in the global arena.* Washington, DC: Climate Institute.

The Nansen Initiative. (2015). *Agenda for the protection of cross-border displaced persons in the context of disasters and climate change.* Vol. 1. https://disasterdisplacement.org/wp-content/uploads/2015/02/PROTECTION-AGENDA-VOLUME-1.pdf

Nash, S. L. (2018). From Cancun to Paris: An era of policy making on climate change and migration. *Global Policy, 9*(1), 53–63. https://doi.org/10.1111/1758-5899.12502

Nash, S. L. (2019). *Negotiating migration in the context of climate change: International policy and discourse.* Bristol, UK: Bristol University Press.

Newland, K. (1979). International migration: The search for work. *Worldwatch Paper, 33,* 1–32.

(NRC/IDMC), N. R. C. D. M. C. (2011). *The Nansen conference: Climate change and displacement in the 21st century.*, June 5–7, 2011. Norwegian Refugee Council/Internal Displacement Monitoring Centre (NRC/IDMC), Oslo, Norway.

Parenti, C. (2011). *Tropic of chaos: Climate change and the new geography of violence.* New York: Nation Books.

Park, L. S.-H., & Pellow, D. (2019). Forum 4: The environmental privilege of borders in the anthropocene. *Mobilities, 14*(3), 395–400. https://doi.org/10.1080/17450101.2019.1601397

Piguet, E. (2010). Linking climate change, environmental degradation, and migration: A methodological overview. *Wiley Interdisciplinary Reviews: Climate Change, 1*(4), 517–524. https://doi.org/10.1002/wcc.54

Piguet, E. (2019). Theories of voluntary and forced migration. In *Routledge handbook of environmental displacement and migration* (pp. 17–28). https://doi.org/10.4324/9781315638843-2

Prieur, M., Marguénaud, J.-P., Monédiaire, G., Bétaille, J., Drobenko, B., Gouguet, J.-J., . . . Shelton, D. (2008). Draft convention on the international status of environmentally-displaced persons. *Revue Européenne de Droit de l'Environnement, 12*(4), 395–406. https://doi.org/10.3406/reden.2008.2058

Rawls, J. (1971). *A theory of justice.* Oxford: Oxford University Press.

Reidhav, D. (2007). *Reasoning by analogy: A study on analogy-based arguments in law.* Lund: Faculty of Law, Lund University.

Renaud, F., Bogardi, J. J., Dun, O., & Warner, K. (2007). Control, adapt or flee how to face environmental migration? Interdisciplinary security connections' publication series of UNU-EHS. Bonn: UNU-EHS. https://reliefweb.int/sites/reliefweb.int/files/resources/F85D742112C97E44C125741900366F86-UNU_may2007.pdf

Rigaud, K. K., Kanta, Sherbinin, A. de, Jones, B., Bergmann, J., Clement, V., . . . Midgley, A. (2018). *Groundswell – Preparing for internal climate migration.* Washington, DC: The World Bank. https://doi.org/doi.org/10.7916/D8Z33FNS

Rodriguez, I. (2020). Latin American decolonial environmental justice. In B. Coolsaet (Ed.), *Environmental justice key issues* (pp. 78–93). Milton: Routledge.

Shacknove, A. E. (2010). Who is a refugee? In H. Lambert (Ed.), *International refugee law* (pp. 163–173). Farnham, Surrey England: Ashgate.

Simonelli, A. C. (2016). *Governing climate induced migration and displacement.* New York: Palgrave Macmillan. https://doi.org/10.1057/9781137538666

Suhrke, A. (1994). Environmental degradation and population flows. *Journal of International Affairs, 47*(2), 473–496.

Suhrke, A., & Visentin, A. (1991, September). The environmental refugee: A new approach. *Ecodecision, 2,* 73–74.

Turhan, E., & Armiero, M. (2019). Of (not) being neighbors: Cities, citizens and climate change in an age of migrations. *Mobilities, 14*(3), 363–374. https://doi.org/10.1080/174 50101.2019.1600913

Unesco de Guchteneire, P., Pécoud, A., & Piguet, E. (2011). *Migration and climate change.* Cambridge: Cambridge University Press.

UNFCCC. (2011). *Decision 1/CP.16. Report of the conference of the parties on its sixteenth session. Part Two.* Cancun, 29 November to 10 December 2010.

UNHCR. (1997). *The state of the world's refugees 1997: A humanitarian agenda.* Oxford: Oxford University Press.

UNHCR. (2009). *Climate change, natural disasters and human displacement: A UNHCR perspective.* Geneva: UNHCR. https://www.unhcr.org/4901e81a4.pdf

UNISDR (United Nations International Strategy for Disaster Reduction). (2015). *Sendai framework for disaster risk reduction 2015–2030.* Geneva: UNISDR.

Vogt, W. (1949). *Road to survival.* London: Victor Gollancz LTD.

Whyte, K., Talley, J., & Gibson, J. (2019). Indigenous mobility traditions, colonialism, and the anthropocene. *Mobilities, 14*(3), 319–335. https://doi.org/10.1080/17450101.2 019.1611015

Wood, W. B. (2001). Ecomigration: Linkages between environmental change and migration. In A. R. Zolberg & P. M. Benda (Eds.), *Global migrants, global refugees: Problems and solutions* (pp. 42–61). New York, NY: Berghahn.

Zetter, R. (2007). More labels, fewer refugees: Remaking the refugee label in an era of globalization. *Journal of Refugee Studies, 20*(2), 172–192. https://doi.org/10.1093/jrs/ fem011

Zickgraf, C. (2019). Keeping people in place: Political factors of (im)mobility and climate change. *Social Sciences, 8*(8), 1–17. https://doi.org/10.3390/socsci8080228

2   The unresolved legal dispute over the recognition of "Climate Refugees"

As observed by Docherty and Giannini, "displacement due to climate change is a de facto problem currently lacking a de jure solution" (Docherty & Giannini, 2009, p. 357). The legal vacuum consists of the lack of a legal instrument specifically governing the issue of "climate refugees." Legally speaking, the major obstacles are that the 1951 Geneva Convention Relating to the Status of Refugees[1] does not include this category of refugees, and no international institution has the mandate to provide international protection and humanitarian assistance to "climate refugees."

But why are climate refugees not recognized as refugees? What are the legal obstacles? What can we learn from the history surrounding the 1951 Convention?

Chapter 2 starts by tracing the origin of the 1951 Convention, setting its geopolitical context, and its minor revisions over the years. It proceeds then by exploring expanded definitions of what constitutes a refugee provided by Regional Refugee Instruments in Africa and Latin America, and offers an overview of different alternatives we may benefit from considering other sources of law such as International Human Rights Law and International Environmental Law.

International refugee law – the history of the 1951 refugee convention

While the practice of seeking asylum is an ancient tradition that goes back to the refugee story of Aeneas, a normative definition of a refugee only appeared with the emergence of nation-states in the nineteenth century (Hurwitz, 2009). Large refugee flows originated by the nation-states system in the first place as it has redesigned the geopolitical order on the ground of ethnonational homogenization, boundary lines redetermination, and armed conflicts that it implies.

In this context, the need to adopt a legal instrument to govern massive refugee exodus first appeared in the aftermath of various conflicts: the Balkan Wars (1912–1913), the World War I (1914–1918), the wars in the Caucasus (1918–1921), and the Greco-Turkish War (1919–1922) (Jaeger, 2001). Tragically, these conflicts caused the breakdown of the centuries-old Russian and Turkish empires, leading to significant numbers of refugees in need of protection.[2]

Given the magnitude of the emergency, the greatest impulse to start the process for setting the international protection system under the leadership of the League

DOI: 10.4324/9781003102632-2

of Nations was given by the Joint Committee of the International Committee of the Red Cross and the League of Red Cross Societies.

On 16 February 1921, such a Joint Committee called a Conference with the principal organizations concerned with emergency relief and the Council. The main issues were the need of more resources to cope with material assistance, and the establishment of a central coordinating body.

To this end, the Council was invited "to appoint a High Commissioner to define the status of refugees, to secure their repatriation or their employment outside Russia, and to coordinate measures for their assistance" (Jaeger, 2001, p. 728).

After preliminary investigation and the proposal being adopted in principle, the President of the Council, Dr. Fridtjof Nansen, accepted to become the first High Commissioner for Refugees of the League of Nations on 1 September 1921. Thus, the history of international protection started with the League of Nations' activities undertaken from 1921 to 1946.

Initially concerned to protect Russian and Armenian refugees, the mandate of the High Commissioner for Refugees was gradually extended to cover Armenians in 1924 up to other categories of refugees such as Assyrians, Assyro-Chaldeans, Syrians, Kurds, and a small group of Turks in 1928 (Jaeger, 2001). Indeed, the first three international arrangements on 5 July 1922, 31 May 1924, and 12 May 1926 were aimed at defining Russian and Armenian refugees in the first place, and dealing with issues of identity certificates and travel documents (Robinson, 1997).

The most relevant arrangement of that phase was the 1928 agreement, a set of resolutions recommending states to adopt measures to protect Russian and Armenian refugees, which ended with the Convention relating to the International Status of Refugees of 28 October 1933.

Considered a milestone in the history of international protection, the 1933 Convention had the merit to install the principle of non-refoulement within an international treaty law and served as a model for the 1951 Convention (Jaeger, 2001). As enshrined by Article 3,

> Each of the Contracting Parties undertakes not to remove or keep from its territory by application of police measures, such as expulsions or non-admittance at the frontier (refoulement), refugees who have been authorized to reside there regularly, unless the said measures are dictated by reasons of national security or public order. It undertakes in any case not to refuse entry to refugees at the frontiers of their countries of origin. It reserves the right to apply such internal measures as it may deem necessary to refugees who, having been expelled for reasons of national security or public order, are unable to leave its territory because they have not received, at their request or through the intervention of institutions dealing with them, the necessary authorizations and visas permitting them to proceed to another country.

Moreover, the 1933 Convention was the first attempt to create a comprehensive legal framework for refugees. It was the first international multilateral treaty to offer Russian, Armenians, and assimilated refugees[3] legal protection and guarantee their basic civil and economic rights. Among the most relevant novelties, the 1933 Convention also introduced administrative measures known as "Nansen certificates" and created committees for refugees responsible for coordinating the work of the organs for finding employment for and assistance to refugees. Unfortunately, the 1933 Convention was significantly limited in its scope to those already considered refugees by the League of Nations, thus failing to assist Jewish refugees fleeing the Third Reich, and was ratified only by nine countries (Jaeger, 2001).

Despite these limitations, the history of international protection under the League of Nations leadership should be seen as a truly innovative and creative period. In addition to the numerous legal provisions discussed earlier, this period also led to the creation of many institutions committed to refugees protection: the Nansen International Office for Refugees (1931–1938), the Office of the High Commissioner for Refugees coming from Germany (1933–1938), the Office of the High Commissioner of the League of Nations for Refugees (1939–1946), and the Intergovernmental Committee on Refugees (1938–1947) (Jaeger, 2001, p. 729). This latter introduced another novelty in the history of international protection by the resolution adopted in Evian on 14 July 1938. For the first time, involuntary emigration from Germany (including Austria) was addressed by extending protection to would-be refugees inside the country of potential departure (Jaeger, 2001). Indeed, the Intergovernmental Committee on Refugees recommended that

> the persons coming within the scope of the activity of the Intergovernmental Committee shall be 1) persons who have not already left their country of origin (Germany, including Austria), but who must emigrate on account of their political opinion, religious beliefs or racial origin, and 2) persons as defined in 1) who have already left their country of origin and who have not yet established themselves permanently elsewhere.[4]

In doing so, the resolution adopted at the Evian Conference provided protection to German refugees who were not included in the previous 1933 Convention, but it left the Jewish question essentially unsolved. After the Nuremberg Laws, German Jews were deprived of their German citizenship, thus becoming stateless refugees. The 1938 Convention failed to protect this group as the only nation that agreed to accept the Jewish refugees fleeing the Third Reich was the Dominican Republic, while other nations did not find any agreement on this issue.

The 1940s were also marked by the proliferation of international bodies, particularly over the post-World War II period (1946–1951) (Hurwitz, 2009). The international cooperation first created the United Nations Relief and Rehabilitation Administration (UNRRA) in 1943. It was a temporary organization aimed at assisting and facilitating repatriations of displaced persons in Europe. These tasks were carried out until its termination in 1947. The responsibility for about 633,000 refugees was then transferred to the International Refugee Organization (IRO), a new organization established on 15 December 1946 by Resolution 62 (I) of the UN General Assembly (Hurwitz, 2009). IRO worked as a resettlement agency providing assistance to refugees and displaced persons, mainly from Central Europe. The beneficiaries were those already protected by the High Commissioner of the League of Nations, the Intergovernmental Committee for Refugees, and the "new" refugees of the period 1947–1950, with the end of World War II. Overall, they amounted to 1.6 million people (Hurwitz, 2009). Unlike the pre-World War II refugees' status, which was regulated by international agreements valid in a restricted number of states, such "new" refugees' status was regulated by IRO agreements with governments. Although the IRO was able to resettle about 1 million refugees (Hurwitz, 2009), successful arrangements were not achieved everywhere, so that new refugees were not treated similarly (Robinson, 1997). In addition, the increasing number of new refugees from Central and Eastern Europe made clear that a resettlement agency designed to end its activities on 30 June 1950, such as the IRO, would not constitute a permanent international instrument for tackling refugees crises.

For these reasons, the Human Rights Commission adopted a resolution for inviting the United Nations to take into due account the legal status of persons who do not enjoy the protection of any government, including the necessary administrative-related issues. Following the resolution, the Economic and Social Council requested the Secretary-General, in consultation with interested commissions and specialized agencies, to undertake a study of the existing situation concerning the protection of stateless persons at both the level of administrative issues and the level of national and international agreements and conventions adopted to date. By the Resolution 116 (VI) of 2 March 1948, the Economic and Social Council also requested the Secretary General of the United Nations to make recommendations on the interim measures which may be taken by the United Nations to further this objective (Robinson, 1997, pp. 3–4).

Thus, the document titled *A Study of Statelessness* marked another milestone in the history of international protection of refugees. The in-depth analysis of the various aspects of the "status of stateless persons" together with the provisions of the 1933 and 1938 Conventions had to constitute a preliminary study before concluding a further convention on this subject. This document urged to take action as the previous Conventions determined the status of stateless persons only for certain groups, while thousands of stateless persons were left without protection. Furthermore, even those protected were at risk as the international organization aimed at facilitating their resettlement, i.e., the IRO, was about to terminate its mandate.

For this reason, the Study suggested appointing an organ responsible for protection within the United Nations Secretariat, a High Commissioner's Office, and continuance of the IRO in another form, or a new specialized agency (Jaeger, 2001). To this end, on 3 December 1949, the UN General Assembly decided to establish a High Commissioner's Office for Refugees whose activities would start from 1 January 1951. Its Statute was adopted by the UN General Assembly on 14 December 1950, when it was also decided to convene a conference of plenipotentiaries in Geneva to draft and sign the Convention relating to the Status of Refugees. Compared to the conferences before the Second World War, this one was attended by representatives of 26 states and 2 observers, thus involving a higher number of states also from other continents than Europe (Robinson, 1997). The process ended on 28 July 1951 with the adoption of the 1951 Convention relating to the Status of Refugees still governing the international refugee protection to date.

Entered into force on 21 April 1954, it provided for a general definition of a refugee and the guarantee of non-refoulement.

Although the higher participation, even from other continents, made the Convention potentially more international and its legal provisions more acceptable to the governments, it remained essentially Eurocentric in its formulation until the 1967 Protocol removed the temporal and geographic limitations initially stipulated in the Convention.

As a post-Second World War instrument, indeed, the scope of the 1951 Refugee Convention was initially restricted to protect persons fleeing events occurring *before 1 January 1951* and *within Europe*. In doing so, those on the move in the aftermath of the post-1945 decolonization process (e.g., 8–10 million people displacement after Pakistan's independence and separation from India) were not covered by the Convention. Such temporal and geographic restrictions were overcome only through the 1967 Protocol, which broadened the scope of the 1951 Convention to *all* persons for events happening at *any* time.

As a result, the legal definition (still nowadays in force) of what constitutes a refugee is to be found in the 1951 Refugee Convention read in conjunction with its 1967 Protocol as a person who:

> owing to well-founded fear of being persecuted for reasons of race, religion, nationality, membership of a particular social group or political opinion, is outside the country of his nationality and is unable, or owing to such fear, is unwilling to avail himself of the protection of that country; or who, not having a nationality and being outside the country of his former habitual residence as a result of such events, is unable or, owing to such fear, is unwilling to return to it.

Even back then, that narrow definition excluded many forced migrants from international protection so that even after the 1951 Convention, other definitions of a refugee in international and domestic laws continued to coexist (Kent & Behrman, 2018). The most relevant example of a refugee definition created outside the 1951 Convention and still in force today is Palestine refugees.

This group-based category of refugees was recognized to respond to the needs of about 750,000 Palestine refugees displaced as a result of the 1948 Arab-Israeli War. To this end, United Nations General Assembly Resolution 302 (IV) of 8 December 1949 established the United Nations Relief and Works Agency (UNRWA) for Palestine Refugees in the Near East. Born in 1950 as a relief and human development agency, even today the UNRWA protects Palestine refugees defined as "persons whose normal place of residence was Palestine during the period 1 June 1946 to 15 May 1948, and who lost both home and means of livelihood as a result of the 1948 conflict."[5] Protection is provided to those meeting this definition, those registered with the Agency, and those in need of assistance, thus including the descendants of Palestine refugee males and adopted children. Although limited in its scope to a specific group, and their descendants, from a particular area at a certain time, this definition is broader than that established in 1951 as asylum seekers are not required to prove they have been persecuted individually (Kent & Behrman, 2018).

Other examples of definitions of a refugee can be found in the Intergovernmental Committee for European Migration, the U.S. "Escapee Program" and the Council of Europe. All these international or governmental organizations used definitions that differed from that adopted by the 1951 Convention. This state of affairs characterized by the coexistence of different definitions of a refugee was opposed by Gerrit Jan van Heuven Goedhart, the first UN High Commissioner for Refugees (Kent & Behrman, 2018). According to van Heuven Goedhart, the coexistence of many definitions of a refugee might lead to an unequal support and levels of assistance provided, while one of the primary aims of the Convention was precisely to define a refugee in uniform terms and avoid the referred disparity.

As noted by Behrman and Kent, however, this critical observation did not take into due account the refugee perspective. Given the narrow definition provided by the 1951 Convention, the existence of organizations other than UNHCR offering protection based on a different definition was better than having no protection at all. In light of this, the authors argue that the consequence of such UNHCR's relatively successful monopolization of the refugee question was that *de facto refugees*[6] were excluded from international protection (Kent & Behrman, 2018, p. 48).

So why did the international community opt for the narrow definition established by the 1951 Convention? What were the drivers that pushed the international community to adopt this legal instrument for tackling refugee crises?

The main reason for adopting an international legally binding instrument such as the 1951 Convention was that the international community recognized refugees as a structural and global phenomenon. Half- or piecemeal measures could work for dealing with a temporary phenomenon, but it was not the case (Robinson, 1997). Further, the magnitude and the global scale of the phenomenon required a concerted effort by all states involved. Therefore, an international legally binding treaty involving many states best suited the need to ensure refugees' fundamental rights and freedoms in uniform terms. By adopting the 1951 Convention and its 1967 Protocol, the international community recognized first its collective responsibility to protect refugees (Hurwitz, 2009), second the need to provide them the same fundamental rights standards and assistance, and finally, the need to agree upon a common working definition, thus overcoming the fragmentation resulting in numerous legal instruments in favor of one legally binding instrument involving the largest number of states. A narrow definition was then the price to pay to get as many states on board as possible.

Among the drivers pushing the international community to do so, it is worth mentioning the idea that an appropriate solution to the refugee problem was an indispensable element in maintaining peace and stability. The aim was to prevent the refugee crisis from becoming a cause of tension between states (Robinson, 1997). A further driver was the necessity to remodel the existing conventions on the basis of the post-war context and rethinking the international law under the United Nations leadership.

At the time of writing, one may conclude that the aims of the 1951 Convention and its 1967 Protocol were only partially achieved, while new challenges are left essentially unaddressed.

On the one hand, as reaffirmed by the more recent international agreement such as the 2018 Global Compact, this international legally binding instrument has succeeded in providing a common working definition of a refugee. In doing so, it has achieved the goal of ensuring more fairness by equalizing the refugees' status with that of nationals of the country of refuge rather than according refugees the most favorable treatment granted to foreigners as it was in the earlier Conventions (Robinson, 1997).

On the other hand, however, it puts the onus on each asylum seeker to prove they have been persecuted, thus rendering more difficult for them to gain protection (Kent & Behrman, 2018).

A further obstacle for protection being driven stems from its overly state-centric approach. Rather than focusing on human beings and their human rights at risk, the 1951 Convention and its 1967 Protocol seem more concerned with assessing the sending state's capacity to ensure its citizens' basic rights and, eventually, the receiving state's obligation to provide protection (Goodwin-Gill & McAdam, 2007, p. 140).

To conclude, compared to the earlier 1933 and 1938 Conventions' approach to refugee definition based on flexible or open groups and categories, the present Convention has moved to a more state-centric and legalistic approach that neglects the refugee perspective and leaves many *de facto refugees* outside the international refugee protection framework (Goodwin-Gill & McAdam, 2007, p. 19).[7]

This outcome is ever more visible at present, marked by a very different geopolitical context where involuntary migration is no longer set in the sole context of wars but increasingly influenced by climate change. In such a new context, it is not surprising that a Convention written 70 years ago offers little if any room for including people fleeing environmental disruptions within its scope.

In confronting "climate refugees" with the refugee definition, scholars have faced various obstacles.

The first challenge is linked to the persecution requirement. Even though some authors have argued either that people fleeing the impacts of climate change ultimately suffer environmental persecution (Conisbee & Simms, 2003),[8] or that climate change is a (new) form of political persecution (Gemenne, 2017), these arguments might be misleading. Not only is the actor of environmental persecution difficult to identify and link with the territory from which the flight occurs, but environmental disruptions are not included among the reasons for persecution set by the refugee definition (McAdam, 2012). Further, asylum seekers have to prove they have been persecuted at the individual level, while the impacts of climate change are more likely to hit large groups of the population.

A second obstacle has to do with the alienage requirement. As it implies that at-risk people have to be outside their own country, it remains to be clarified if and to what extent this requirement applies to "climate refugees" who rarely cross borders, becoming more often internally displaced people (Ionesco, 2017; McAdam, 2012; Rigaud et al., 2018). Internally displaced people (IDPs) constitute a distinct category introduced by the 1998 Guiding Principles on Internal Displacement: a non-legally binding instrument established to protect those forced to flee but who do not cross international borders. As laid down by Article 2,

> internally displaced persons are persons or groups of persons who have been forced or obliged to flee or to leave their homes or places of habitual residence, in particular as a result of or in order to avoid the effects of armed conflict, situations of generalized violence, violations of human rights or natural or human-made disasters, and who have not crossed an internationally recognized State border.[9]

At the time of writing, the only existing legally binding instrument to protect IDPs is the 2009 African Union Convention for the Protection and Assistance of Internally Displaced Persons In Africa. The so-called Kampala Convention explicitly protects and assists "those who have been internally displaced due to natural or human made disasters, including climate change."[10] As a result, those people are protected only in the African States that ratified the Convention.

Over the years, critical scholars fueled a lively debate to overcome these obstacles. In particular, some objected to the narrow interpretation given to the term "persecution" in the first place (Andrade, 2008; Coles, 1990; Cooper, 1997; Gemenne, 2017; Hathaway, 2017; Shacknove, 1985). First, this term has not been codified under International Refugee Law because of the difficulty of listing all possible forms of persecution that enable international protection (Hathaway, 2017). For this reason, drafters of the 1951 Convention preferred leaving the term undefined and open to broader interpretation. Second, a doctrinal endeavor pursued by the United Nations High Commissioner for Refugees (UNHCR) over time has clarified several forms of harm that amount to persecution (Andrade, 2008). These include:

(i)   serious physical harm, loss of freedom, and other serious violations of basic human rights as defined by international human rights instruments;
(ii) discriminatory treatments which lead to consequences of a substantially prejudicial nature (for instance, serious restriction on the applicant's right to earn his or her living, to practice his or her religion, to access normally available education facilities); and
(iii) a combination of numerous harms none of which alone constitutes persecution but which, when considered in the context of a general atmosphere in the applicant's country, produces a cumulative effect that creates a well-founded fear of persecution. (Andrade, 2008, p. 124).

In this view, an "environmental persecution" would seem conceivable under the third form of harm. Further, as recalled by De Andrade, according to the case law, perpetrators of persecution are no longer restricted to state actors and the intentionality of persecution is no longer a requirement (Andrade, 2008). The same *ratio* applies to these developments in case law: the focus shifts from the agent of persecution to the victims of human rights violations. Such shift is particularly sound in contexts such as Somalia, Afghanistan, and the like, where sovereignty is at stake and persecution is increasingly emanating from non-state agents.[11]

Most academic interpretations of persecution also suggest understanding the term as the failure of state protection resulting in systemic human rights violations (Hathaway, 2017).

This interpretation was also upheld by Shacknove, who argued that neither persecution nor alienage is a necessary condition for establishing refugee status (Shacknove, 1985). While persecution is only one manifestation of the lack of state protection of the citizens' basic needs, alienage should be interpreted as a subset of a broader category: access to international protection. In Shacknove's perspective, people in vulnerable situations must not cross the border to get protection from the international community. In a similar line of reasoning, Coles also criticized the persecution-based standard of refugee status on the ground that persecution might also be suffered by citizens still inside their home States (Coles, 1990). Like Shacknove, Coles also argued the need to rethink the criteria of refugeehood toward a closer emphasis on the coercive nature of the refugee's separation from her home community (Coles, 1990).

However, despite scholars' efforts and recent developments, persecution and alienage are still evoked as legal challenges. In light of this, the work in this chapter suggests that the 1951 refugee definition keeps on leaving many *de facto refugees* outside the international refugee protection framework. Thus, given its limits in the face of global challenges such as climate change, are there any better alternatives?

The next section proceeds by analyzing other existing refugee definitions included in two regional refugee instruments in force in Africa and Latin America: the 1969 OAU Convention and the 1984 Cartagena Declaration.

Regional refugee instruments: OAU convention and the Cartagena declaration

Although the 1967 Protocol had helped broaden the scope of international refugee protection by removing temporal and geographic limitations, individual persecution still constituted a significant constraint. People fleeing armed conflicts or serious public disorder, ever more frequent at the time of decolonization, still fell outside the refugee definition established by the 1951 Convention and its 1967 Protocol. A major obstacle was the onus on each asylum seeker to justify fear of persecution in certain large-scale displacement situations as those experienced at that time in Africa. Thus, even after the 1967 Protocol came into the picture of international refugee protection, challenges arising from different contexts clarified the urgency to expand the refugee definition further.

In this regard, the 1969 Convention governing the Specific Aspects of Refugee Problems in Africa (hereinafter OAU Convention) may be well considered the first departure from the universal approach adopted in Geneva.

Since the 1960s, decolonization campaigns were still ongoing, and the political situation in Africa was marked by massive violence. Many freedom fighters fueled struggles in African states remained under colonial or white minority rule, while African states that recently gained independence experienced or risked being involved in civil wars due to attempts to bring about territorial readjustments by postcolonial states or other political actors.

As many people forced to flee from violence and conflicts did not find protection in the 1951 Convention due to its temporal and geographic limitations, the OAU's Council of Ministers formed a commission of representatives from 10 OAU member states in 1964. The reason for drafting a regional refugee treaty such as the 1969 OAU Refugee Convention was initially to make international refugee law applicable in Africa. When the Protocol made the international refugee protection universally applicable in 1967, African states questioned the necessity of having a regional instrument. Following discussions and debates during the Conference on the Legal, Economic and Social Aspects of African Refugee Problems held in Addis Ababa on 9–18 October 1967, the conclusion was reached that the "1951 Convention definition – while universally applicable – might not be sufficient to cover all refugee situations in Africa" (Jackson, 1999, p. 187). The focus was then shifted from making the international refugee protection applicable in Africa to addressing refugee issues specific to Africa (Abebe, Abebe, & Sharpe, 2019).

The first reason for considering the 1951 Convention and its 1967 Protocol inadequate to address specific context-related refugee issues in Africa was the lack of distinction between African refugees from independent African states and those from countries struggling against colonial or white minority rule. In the absence of such a distinction, the 1951 Convention would not have prevented refugees from using countries of asylum as bases from which to overthrow the regimes in their countries of origin, and, what is worse, would not have qualified freedom fighters as refugees (Abebe et al., 2019). Furthermore, the fear of persecution requirement to be met by individuals did not fit large-scale influx situations that frequently occurred in Africa at that time. As a result, all these context-specific issues led to the OAU Convention adoption in Addis Ababa on 10 September 1969. (Then solved by Article III of OAU Convention that enshrined the prohibition of subversive activities.)

The shortcomings identified in the 1951 Convention's universal approach were then addressed by the prohibition of subversion provided for in Article 3 and the broadened definition of a refugee enshrined by Article 1(2) of the OAU Convention.

As recalled in the Preamble, "anxious to make a distinction between a refugee who seeks a peaceful and normal life and a person fleeing his country for the sole purpose of fomenting subversion from outside," heads of State and Government convened in Addis Ababa in 1969 agreed on prohibiting subversive activities. To this end, Article 3(1) establishes that every refugee shall abstain from any subversive activities against any Member State of the OAU. Also, Article 3(2) further specifies that "signatory States undertake to prohibit refugees residing in their respective territories from attacking any State Member of the OAU."

To facilitate the recognition of refugee status to individuals fleeing armed conflicts and violence in large-scale influx situations, Article 1(2) lays down that

> The term 'refugee' shall also apply to every person who, owing to external aggression, occupation, foreign domination or events seriously disturbing public order in either part or the whole of his country of origin or nationality, is compelled to leave his place of habitual residence in order to seek refuge in another place outside his country of origin or nationality.
>
> (OAU Convention, Art.1, 2).

It is worth noting that such an extension is not to be considered exclusive but rather overlapping and complementary to that established by the 1951 Convention and its 1967 Protocol. Its complementary nature can be well inferred from the lack of a refugee rights framework. The reason for this omission is that refugees under the OAU Convention can derive these rights from Articles 3–34 of the 1951 Convention and human rights law (Abebe et al., 2019). Indeed, the primary goal of the OAU Convention was to extend the refugee definition to a wider range of individuals, thus acting as an effective regional complement.

Its broadened definition remains relevant even today as it allows to protect individuals fleeing conflicts and massive violence more effectively. Thanks to Article 1(2) and its "events seriously disturbing public order" clause, those individuals can benefit from a prima facie approach to refugee status determination allowing members of large groups of displaced to be considered individually as refugees. Further, unlike the 1951 Convention definition, its applicability is dispensed by the need to prove they have been persecuted (Jackson, 1999).

Another novelty that rendered the OAU Convention tailored to the African cultural context was the international cooperation approach driven by the spirit of African solidarity aimed at alleviating the burden of the Member State granting asylum. In this regard, Article 2(4) sets out the following modalities for regional responsibility sharing:

> where a Member State finds difficulty in continuing to grant asylum to refugees, such Member State may appeal directly to other Member States and through the OAU, and such other Member States shall in the spirit of African solidarity and international co-operation take appropriate measures to lighten the burden of the Member State granting asylum.[12]

This provision aimed at alleviating the burden of states most impacted by refugee flows through a spirit of solidarity among states was underestimated in the 1951 Convention. Significantly, this question of fairness has been fully integrated only in the 2018 Global Compact on Refugees by the burden and responsibility-sharing principle (see next section in this chapter).

Entered into force on 20 June 1974, the OAU Convention had a significant influence even outside Africa.[13] In particular, it influenced UNHCR's mandate and informed regional instruments such as the Cartagena Declaration in Latin America.

As for the UNHCR's mandate, in 1981, the UNHCR Executive Committee recommended extending the 1951 Convention refugee definition to mass displacement by using the wording of the 1969 OAU Convention (Abebe et al., 2019). Thus, since 1981, refugee status determinations are conducted by taking into due consideration the "events seriously disturbing public order" clause provided by Article 1(2) of the OAU Convention.

As reaffirmed in the revised and updated 2020 version of the Procedural Standards for RSD (Refugee Status Determination) under UNHCR's Mandate (RSD Procedural Standards),[14] even today, UNHCR's broader refugee criteria include a situation of generalized violence or events seriously disturbing public order. These situations can lead to a prima facie approach to refugee status determination.

> A prima facie approach means the recognition of refugee status on the basis of readily apparent, objective circumstances in the country of origin (or, in the case of stateless asylum-seekers, their country of former habitual residence) indicating that individuals fleeing these circumstances are at risk of harm which brings them within the applicable refugee definition, rather than through an individual assessment.
>
> (UNHCR, 2020, p. 112)

As for the 1984 Cartagena Declaration in Latin America, a non-binding instrument adopted by the Colloquium on the International Protection of Refugees in Latin America, Mexico, and Panama, its broadened refugee definition has included "persons who have fled their country because their lives, safety or freedom have been threatened by generalized violence, foreign aggression, internal conflicts, massive violation of human rights or other circumstances which have seriously disturbed public order" (Cartagena Declaration, III, 3). This definition clearly reflects the wording of Article 1(2) provided for the 1969 OAU Conventions, as well as its complementary nature. Indeed, even this declaration highlights that the definition has to be meant "in addition to containing the elements of the 1951 Convention and the 1967 Protocol" (Cartagena Declaration, III, 3).

Given the then existing 1967 Protocol and the 1981 prima facie approach to refugee status determination, one may well ask why a regional refugee instrument was considered necessary in Latin America. Yet, the reasons for adopting a regional conceptual framework for refugee protection policy rely on Latin America's historical context in the late 1970s and early 1980s.

Over that decade, it is estimated that various conflicts occurred in Honduras, El Salvador, Nicaragua, and Guatemala displaced around two million people (Fischel De Andrade, 2019). Triggered by anti-communist policies of different United States (US) Administrations due to the Cold War's geopolitics, those conflicts generated large-scale forced migrants influxes, including forced migrants who did not fall under the 1951 refugee definition (Fischel De Andrade, 2019).

A further cause of displacement in Nicaragua was the devastating earthquake in 1972 that destroyed nearly 90 percent of its capital, Managua. It provoked large-scale displacement, above all, due to the resulting food shortage suffered by citizens.

In the face of these challenges, Central America was not equipped with coherent and adequate policies to respond asylum-seekers' needs of protection. However, the urgency to address massive forced migration was not the only driver for triggering such a change regime. Even more important in the Latin American context was the UNHCR's protection advocacy work.

In UNHCR's view, its own 1981 broad definition extending protection to mass displacement was not sufficient to cope with large-scale group situations occurred in Latin America. Indeed, only few states were parties of the 1951 Refugee Convention and could not address the magnitude of that large-scale forced migration. Thus, the advantage of adopting a wider definition in the whole region would help in granting refugee status to those fleeing from consequences of armed conflicts and violence without creating another category of protected persons.

The 1981–1984 UNHCR-led events to advance such a regional conceptual framework for refugee protection policy culminated with the 1984 Cartagena Declaration adoption. It provided a broader definition able to ensure forced migrants the same rights and duties enjoyed by 1951 Convention refugees, thus avoiding creating a parallel legal status in Latin America.

Although not legally binding, the Cartagena Declaration represents a common instrument of policy and guiding principles for the whole region and had a significant influence on region's domestic legislation. Even though many states have transposed the wider Cartagena definition in different ways, thus rendering difficult to identify a common, regional refugee definition *stricto sensu*, the 1984 Cartagena Declaration has still achieved its aim to grant refugee status to forced migrants fleeing the consequences of armed conflict and violence (Fischel De Andrade, 2019).

Having accomplished the history of the broaden definitions provided by the OAU Convention and the 1984 Cartagena Declaration, the main question is whether those regional instruments may eventually be used to include "climate refugees" in the refugee protection spectrum.

The previous section ended concluding that the 1951 Refugee Convention was, ultimately, a product of its time. Can we say the same for the OAU Convention and the 1984 Cartagena Declaration? Are they better equipped to cope with new challenges arising from the impacts of climate change?

Notably, Africa and Latin America are among the regions most affected by environmental degradation. Even though its impacts are becoming ever more visible at the time of this writing, the emergence of such slow environmental violence is not "new." Rather, it runs in parallel with the violence of wars occurred in those contexts since the 1960s.

Ever since then, environmental issues have also played a significant role in exacerbating involuntary migration. Examples include the series of droughts, which began in 1968 and spread from the Sahel to southern Africa, forcing Sahelians to move south and west to the coastal West African nations, and floods affecting 221 million people worldwide between the 1960s and 1980s (El-Hinnawi,

1985). Particularly disruptive were floods which occurred in 1983. To name but a few, floods and landslides in Ecuador caused US$400 million worth of crop and property damage; record floods severely affected eastern lowlands in Bolivia; and in Peru, the government declared a state of emergency in response to epidemics of typhoid and dysentery in four northern provinces because of the heavy rains (El-Hinnawi, 1985, pp. 14–15).

The situation is even worse today. In 2020, new displacements amounted to ca. 40.5 million, with disasters responsible for over three times compared to conflicts and weather-related events causing 98 percent of all disaster displacement (IDMC, 2021). For this reason, confronting the OAU Convention and the 1984 Cartagena Declaration with environmental challenges is ever more necessary today.

At first glance, both regional refugee definitions have a clear advantage compared to the one established by the 1951 Convention, i.e., the "events seriously disturbing public order" clause.

On paper, environmental disasters might be well conceived events able to disturb public order, but in concrete the issue is not straightforward.

The first challenge stems from the fact that both the OAU Convention and the 1984 Cartagena Declaration require evidence of an *actual* threat able to disturb public order. Unlike the 1951 Convention, including the risk of *potential* future harm, the requirement of such an *actual* threat does not apply to the new displacements that mostly take place in the form of pre-emptive evacuations (McAdam, 2012).

The second challenge is linked to the political impacts that environmental disasters should provoke.

Indeed, such events rarely result in a breakdown in law and order, so that most of them ultimately remain outside the scope of the legal instruments at stake (Mcadam, Burson, Kälin, & Weerasinghe, 2016).

In light of these obstacles, it has remained unclear whether environmental disruptions may serve as readily apparent, objective circumstances in the country of origin leading to the mass displacement, thus triggering the *prima facie* approach. If those were conceived as objective circumstances, the obstacles deriving from the persecution requirement – such as the onus of proof on asylum seekers and refugee status determined at the individual level – would be easily overcome.

Although debates are still ongoing and the issue remains unsettled, the history of these wider refugee definitions allows us to conclude that (i) the 1951 refugee definition has been already revised in the past and (ii) changes in the refugee concept's understanding have occurred in response to changes in the geopolitical context.

Toward the global compacts on refugees and migration and beyond

Since the OAU Convention and the Cartagena Declaration have introduced broader refugee definitions, changes in the geopolitical context have further occurred. The breakup of the Soviet Union, the end of the Cold war era, and China's entry into the global economy have posed legitimate questions and hopes concerning the rising of the new world order. Options discussed by scholars were whether:

1   it was possible to re-shape the global system by concerted states' efforts through the United Nations;
2   it was the dawn of a new geopolitical context under the US global hegemony and its pillars based on Western liberal democracy and the free market; and
3   for a new world order to be built, most prominent power centers would have to take collective action to stabilize and enhance the international system.

At the time of writing, various factors point to the conclusion that the third option is more likely to materialize. The rise of new power centers (EU, Japan, an invigorated Russia under Putin's leadership, a rising China, and emergent India and Brazil), and new global threats (such as climate change and international terrorism) have ignited hopes that a new world order will be based on collective efforts to maintain peace and enhance the international system.

While it is still unclear which direction the US foreign policy will take under the Biden Administration, the new US President has re-joined the Paris Agreement, thus confirming a visible convergence of major power centers on fighting a global threat such as the climate change.

Adopted by 196 Parties at COP 21 in Paris on December 12, 2015, the Paris Agreement is considered a historic agreement for climate activists, being the first legally binding international treaty on climate change bringing all nations into a common cause, including the major CO_2 emitters in the world: China and the US.

When Former President Trump announced in June 2017 and formally withdrew the US from the Paris Agreement on November 4, 2019, heads of state, business leaders, and environmentalists worldwide expressed their disappointment with the decision. Above all, they expressed serious concerns that the Paris Agreement's goal to limit global warming to well below 2, preferably to 1.5 degrees Celsius, compared to pre-industrial levels, could have not been met without the US commitment. Thus, the US's re-engagement confirms that the fight against climate change is a topic enabling the convergence of the major power centers on a common goal. Of utmost importance, the 2021 virtual *Leaders Summit on Climate*, convened by President Biden, further reaffirmed this core idea and served as an opportunity for Biden to assure other countries of the US' willingness to take bolder action and leadership on climate change.[15] The summit was the largest virtual gathering of its kind, with about 40 world leaders involved. During this summit, President Biden – along with John Kerry and Antony Blinken, US climate envoy and secretary of state, respectively – announced their commitment to pursue ambitious climate targets set by the Clean Future Act. Tabled in the House of Representatives in early March 2021, the Clean Future Act set the US ambitious goal of becoming a net-zero greenhouse gas economy by 2050. Similarly, China's President Xi Jinping also announced his willingness to achieve climate neutrality by 2060. Despite China's strong dependence on coal,[16] President Xi Jinping promised for the first time to phase out coal as an energy source from 2025.

Consistent with its advanced policies concerning environmental sustainability, the EU aligned its climate goals toward the zero emissions target of greenhouse gases by 2050. To achieve this goal, the EU adopted the European Green Deal: a set of policy initiatives funded with one third of the 1.8 trillion euro investments from the *NextGenerationEU Recovery Plan* introduced in response to the Covid-19 crisis and the EU's seven-year budget.

Thus, the US, China, and the EU have established themselves as central players in the race for climate neutrality and energy transition as common goals. In particular, the post-pandemic recovery plans have been adopted with the main purpose of transforming their energy systems while competing for leadership in the future geopolitical and geo-economic scenario.[17]

Such common goals are also in line with the 2030 Agenda for Sustainable Development adopted by all United Nations Member States in 2015 to tackle climate change and environmental protection, thus facilitating the convergence ever further. In this regard, 2015 can be well considered a turning point, being marked by multilateralism and numerous international agreements on climate change-related issues. In addition to the Paris Agreement and the 2030 Agenda for Sustainable Development, the following agreements and initiatives have been adopted in 2015:

- Sendai Framework for Disaster Risk Reduction 2015–2030
- Nansen Initiative's agenda for protection (that gave rise to the Platform on Disaster Displacement, PDD).
- International Organization for Migration (IOM) Division on Migration, Environment and Climate Change (MECC)
- UNFCCC-led Task Force on Displacement

What all these initiatives have in common is the focus on human mobility in the context of climate change and disasters. Adopted at the Third UN World Conference in Sendai, Japan, on March 18, 2015, the Sendai Framework for Disaster Risk Reduction is the successor instrument to the Hyogo Framework for Action (HFA) 2005–2015: Building the Resilience of Nations and Communities to Disasters. A significant novelty compared to HFA has been the shift from a disaster management to a disaster risk management approach. As established by the Sendai Framework para 19(a), "each State has the primary responsibility to prevent and reduce disaster risk." This emphasis on disasters prevention in the first place for reducing the risk disasters may happen is due to the adverse impacts they have on people living in the areas concerned. Over the last decade, the number of people forced to move internally or abroad to find refuge in the aftermath of disasters or climate-change-related phenomena has increased up to the form of large-scale displacement. However, the Sendai framework refers only to temporary movements of people without providing any durable solution such as permanent relocations, protracted displacement or circular migrations.

To cope with these challenges, three main initiatives have been advanced so far: the Nansen Initiative, the International Organization for Migration (IOM)

Division on Migration, Environment and Climate Change (MECC), and the UNF-CCC-led Task Force on Displacement.

The Nansen Initiative is a state-led initiative launched by Switzerland and Norway in 2012 with the aim of protecting people displaced across borders in the context of disasters and climate change (see also Chapter 3 of this volume). The initiative gave rise to the Protection Agenda and the Platform on Disaster Displacement (PDD). The endorsement of the Protection Agenda to fill the protection gap concerning people on the move in the context of disasters and climate change determined the end of the Nansen Initiative in October 2015 and the start of the new initiative, i.e., the Platform on Disaster Displacement, on May 23, 2016. The primary aim of the platform is, indeed, to implement the recommendations included in the Protection Agenda. Above all, one of the main strategic priorities is promoting effective practices to prevent and reduce displacement through national measures that reduce disaster and displacement risk combined with climate change adaptation plans.

Established in 2015, the International Organization for Migration (IOM) Division on Migration, Environment and Climate Change (MECC) has been the first institutional setting entirely dedicated to human mobility climate-related issues. Since the year of its creation, the MECC has strengthened the cooperation on cross-cutting issues with all the IOM's departments (especially with IOM gender unit), while collaborating with external actors such as other intergovernmental organizations (IGOs), non-governmental organizations (NGOs), and academics. Among its main priorities, the MECC has worked hard to promote advocacy, support capacity building, invest in data collection, build environmental migration policies with a particular focus on vulnerable populations, and bolster environmental sustainability. Thanks to the IOM Strategic Vision 2019–2023, the MECC has also been tasked to develop the IOM Wide Institutional Strategy on Migration, Climate Change and Environment. Launched in 2019, this institutional strategy is based on three pillars: resilience, mobility, and governance. Its main goal is to lead global coordination on (environmental) migration issues while integrating long-term climate and environmental perspectives in its policies. MECC's institutional engagement envisioned for the near future will be based on existing knowledge accumulated by IOM over the last 20 years.

The UNFCCC-led Task Force on Displacement, instead, has been set up under the Paris Agreement in 2015. Indeed, the COP 21 requested the Executive Committee of the Warsaw International Mechanism (WIM) to establish this task force for recommending integrated approaches to avert, minimize, and address displacement related to climate change's adverse impacts. The task force consists of 14 members representing the most relevant actors working in climate migration/displacement field including intergovernmental organizations (IGOs), non-governmental organizations (NGOs), international humanitarian organizations, existing bodies and experts groups (including the Adaptation Committee and the Least Developed Countries Expert Group), civil society, the Platform on Disaster Displacement (PDD), and Executive Committee of the Warsaw International Mechanism for Loss and Damage. The task force started its work in June 2017, and issued its first set of recommendations in 2018.

Given the similar nature of their compositions, the recent literature has described the UNFCCC Task Force on Displacement and IOM's MECC Division as emerging models of cross-governance: a "formal, accommodating institutional setting, in which actors from different institutions, areas of expertise and perspective can operate (and cooperate) in a coordinated manner" (Kent & Behrman, 2018, p. 152). Both institutional settings have been constituted to govern this complex policy making area, at the intersection between climate change and migration governance, that has long been characterized by institutional fragmentation and lack of coordination of the numerous institutional actors involved (Martin, 2010). Thus, the MECC Division at IOM and the UNFCCC TDF represent the most relevant examples of institutional structures addressing the "institutional gap" that surrounds the international governance of climate migration/displacement. By the expression "institutional gap," I am referring to the fact that no international institution has the explicit mandate to provide international protection and humanitarian assistance to people forced to migrate because of the impacts of climate change. Nor do international agencies, organizations, and relevant actors working in this field have established mechanisms for creating an overarching coordinating framework.

Not surprisingly, all these agreements were adopted in 2015, considered the year of the global refugee crisis. Not only the Syrian civil war, which some have portrayed as linked to climate change (Kelley, Mohtadi, Cane, Seager, & Kushnir, 2015), has sparked the refugee crisis in Europe, but also other conflicts such as those in Afghanistan, Iraq, and Libya have triggered unprecedented refugee and migrants flows. In particular, the main routes to Europe included the Mediterranean, with people fleeing war and violence from Libya to Italy, and the route from Turkey to Greek islands, such as Lesvos. Above all, the tragedy occurred in April 2015, when over 600 people drowned in the Mediterranean some 180 kilometers south of Italy's Lampedusa Island, served to initiate a process for rethinking the refugee/migration regime in light of the growing number of people on the move fleeing wars, conflicts, and climate change's impacts.

In this context, the first step was fostering international community's greater responsiveness to large movements of refugees and migrants. In particular, the primary aim was to improve the situation of refugees and migrants on the move by laying the groundwork for the development of two Global Compacts, one on refugees and the other on migrants. Following the Summit for Refugees and Migrants hosted on 19 September 2016 by the United Nations General Assembly, all 193 Member States of the United Nations unanimously adopted the New York Declaration for Refugees and Migrants (Resolution 71/1), which is considered a milestone for protecting those who are forced to flee, and easing pressure on host countries. Further, the 2016 New York Declaration for Refugees and Migrants explicitly recognizes that the increasing number of people fleeing wars, violence, human rights' violation, natural disasters, and climate change's impacts calls for global and durable solutions.

Thus, such a process toward concerted efforts to tackle the diverse and context-specific challenges of people on the move due to wars, persecutions, natural disasters, climate change's impacts, or a combination of these reasons culminated with the Global Compact for Safe, Orderly and Regular Migration (GCM) and Global Compact on Refugees (GCR) adoption in 2018. These international agreements recognize the need for cross-sectoral, coordinated action to address this pressing issue.

The novelties of these two non-legally binding instruments rely, on the one hand, on the acknowledgment of natural disasters, the adverse effects of climate change, and environmental degradation as drivers of refugees/migrants movements. On the other hand, they rely on the similar vulnerabilities migrants and refugees face in large-scale influx, including the same universal human rights and fundamental freedoms they are entitled to.

The Global Compact on Refugees has better addressed the question of fairness (Cantor, 2019) between developed and developing or least developed states concerning the equitable, predictable, and sustainable sharing of the burden and responsibility resulting from the presence of large numbers of refugees (Part A). Indeed, the principle of burden- and responsibility-sharing aims to alleviate the burden on host countries and communities affected by large-scale refugee movements. Ultimately, the Global Compact on Refugees aims to prevent displacement, increase responsiveness, and find durable solutions for displaced persons (McAdam, 2019). Above all, those with diverse needs and potential vulnerabilities (Part B).

By contrast, the Global Compact for Safe, Orderly and Regular Migration has introduced a holistic approach aimed at addressing migration in all its dimensions. This approach implies to promote multi-stakeholder partnerships for enhancing and strengthening migration governance. As a result, the following actors are involved to ensure horizontal and vertical policy coherence across all sectors and levels of government: migrants, diasporas, local communities, civil society, academia, the private sector, parliamentarians, trade unions, National Human Rights Institutions, the media and other relevant stakeholders in migration governance.[18]

Although on paper the Global Compacts rest on the Paris Agreement, the 2030 Agenda for Sustainable Development, and the Sendai Framework for Disaster Risk Reduction, a closer inspection reveals that these instruments did not cope with creating new legal subjectivities such as climate refugees/migrants (Bufalini, 2019). By contrast, as explicitly mentioned in the Global Compact on Migration's Preamble, migrants and refugees are two distinct categories governed by different legal frameworks. Thus, the Global Compact for Migration, on the one hand, and the Global Compact on refugees, on the other hand, reaffirm the dichotomy between refugees and migrants, with only refugees being entitled to the specific international protection as established by international refugee law. Nor has the definition of a refugee been rethought or amended to include climate refugees.

The Global Compact for refugees is grounded in the international refugee protection regime, i.e., the 1951 Refugee Convention and its 1967 Protocol, and other regional refugee instruments. This means that such a new instrument does not introduce any broader definition of a refugee nor provides for new legal norms. Rather, it only sets the framework for applying existing norms in large-scale refugee flows in the light of the innovative burden- and responsibility-sharing principles. Similarly, the Global Compact for Migration is imbued with references to climate change mitigation and adaptation and sustainable solutions.[19] However, in the face of irregular migration, in which most climate refugees/migrants end up being classified, it has not created any new category of climate migrants, nor has it provided for new, safe, and legal routes of migration. On top of that, it has not set any specific targets by which to increase legal routes, thus confirming that its main goal is more likely to prevent irregular migration in the first place (Appleby & Kerwin, 2018). To this end, indeed, the Global Compact for Migration has envisioned to increase international and regional cooperation to promote the implementation of the Agenda 2030 for Sustainable Development in geographic areas from where irregular migration mainly originates rather than creating new legal routes of migration.[20]

Ultimately, the Global Compacts have added more soft law and general principles without rethinking the meaning and governing of refugee/migration regimes. In particular, they seem to reflect and conserve the contradictions experienced during the negotiations between the "conservative wing," more concerned to reaffirm a state-centric approach focused on the primacy of national policies and priorities, and the "progressive wing," more inclined to a multi-stakeholder approach for promoting mutual cooperation in the spirit of solidarity. It is worth noting that the Global Compact for Migration was drafted by States, while the drafting of the Global Compact on Refugees was led by a UNHCR still constrained by what States would agree to (McAdam, 2019).

Thus, these two wings seem to coexist in both instruments without having solved the underlying contradictions.

Despite the novelties introduced by the Global Compact on refugees, it has not changed the more-deep rooted state-centric approach as evidenced by the following, recurrent expressions: "upon the request of concerned host countries or countries of origin"; "in support of host countries and in line with national laws, policies and strategies"; "in close cooperation with national authorities of host countries"; "in support of national development plans and strategies"; "in close coordination with national institutions"; "respecting the primacy of national ownership and leadership"; "in line with country ownership and leadership and respecting national policies and priorities"; "at the request of concerned States, and in full respect of national laws and policies"; and "in line with national laws and policies." All these expressions confirm the utmost importance given to the principle of sovereign equality of States in governing large-scale influx of refugees. Even the most significant novelty of the Global Compact on refugees, i.e., the operationalization of the principle of burden- and responsibility-sharing, ultimately applies to states and aims at easing pressure on developing or least developed states most affected by large refugee movements. Indeed, several aims are devoted to the environmental impact of refugees on host countries.[21]

To conclude, although the Global Compacts acknowledge changes in the geopolitical context and the growing influence of climate change's impacts on human migration, these instruments are not legally binding, and do not introduce new legal categories, nor a legal status for people fleeing environmental disruptions.

Limits and possibilities of the refugee law concepts

A further possibility to include people fleeing climate change's impacts within the "refugee realm" may be found in the non-refoulement principle enshrined in Article 33(1) of the 1951 Refugee Convention. According to Article 33(1),

> No Contracting State shall expel or return ("refouler") a refugee in any manner whatsoever to the frontiers of territories where his life or freedom would be threatened on account of his race, religion, nationality, membership of a particular social group or political opinion.[22]

This principle implies that states are not allowed to repatriate a refugee or asylum seeker to their countries of origin if their life or freedom is at risk or threatened upon return. Against this background, some scholars have questioned whether environmental disruptions are likely to trigger the non-refoulement principle. To date, this lively debate has generated different answers.

Benoît Mayer holds that there are some reasons for caution. First, environmental degradation is not included in the five grounds for persecution, triggering the non-refoulement principle. Given that the wording of the Refugee Convention uses the expression "for reason of," Mayer concludes that utmost importance is devoted to the "discriminatory intent rather than a mere discriminatory effect" environmental disruptions may have (Mayer, 2016, p. 113). Another reason for caution is that evidence shows that people fleeing environmental disruptions rarely cross the border, which is, instead, a necessary requirement of the non-refoulement principle. Finally, recognizing the refugee status of such people on the move would mean accepting the idea that only forced migrants deserve protection. By contrast, this particular case should be used to rethink migration governance by strengthening the human rights protection of all migrants. Not only this approach would allow to reduce migrants' vulnerability exacerbated by climate change's impacts, but it is also more politically feasible than extending the non-refoulement principle.

Other authors have proposed different perspectives. In particular, Michelle Foster (2009) has recalled that there are already cases of people protected by the non-refoulement principle beyond the terms (and wording) of the Refugee Convention. After the UNHCR observed in its 2006 Note on International Protection that

> there may also be persons with international protection needs who are outside the refugee protection framework, requiring finer distinctions to be made to provide protection in ways complementary to the 1951 Convention.[23]

many states have implemented complementary protection by their national legislations. Therefore, the contending issue is not if it is possible to extend the non-refoulement beyond the wording of Refugee Convention, but rather to which extent the scope of protection can be broadened. According to Foster, the point is whether complementary protection may be ensured to people who fear torture or right to life violation only, or it may be ensured also to people suffering socioeconomic rights' deprivation. A person's life may be at risk also due to famine, or lack of medical treatment upon return to his/her country of origin, thus calling into question the possibility to invoke a state's international protection obligations (Foster, 2009). In this context, environmental degradation may well cause famine, or unhealthy living conditions in developing countries less equipped to cope with these issues. However, although the Executive Committee of UNHCR pointed out that such international protection – grounded in international and regional human rights instruments adopted after the Refugee Convention – constitutes a legal obligation rather than a mere discretionary, and charitable decisions of states, many states have considered socio-economic rights' deprivation a matter of humanitarian discretion (Foster, 2009).

A more radical position can be found in Betts' understanding of non-refoulement principle. According to Betts, not only non-refoulement should not be limited to persecution and armed conflicts or extended only to situations related to climate change or environmental disruption, but it should apply indiscriminately to all "survival migrants." Those are "persons who are outside their country of origin because of an existential threat for which they have no access to a domestic remedy or resolution" (Betts, 2016, p. 278). Following his definition, the prominent issue is setting the threshold for identifying an existential threat in the country of origin. To this end, Betts draws on the concept of "basic rights" developed by Henry Shue (1980), and conceived as "the basic conditions for anyone to enjoy any other right" (Betts, 2016, p. 24). Such an ethical perspective has been rejected by Mayer, who has criticized Betts for not defining sufficiently the just threshold or nature of coercion justifying protection (Mayer, 2016).

A useful indication to better understand the threshold justifying international protection can be found in the HRC General Comment 31 (2004) on the Nature of the General Legal Obligation on States Parties to the Covenant. As stated in para 12,

> the article 2 [ICCPR] obligation requiring that States Parties respect and ensure the Covenant rights for all persons in their territory and all persons under their control entails an obligation not to extradite, deport, expel or otherwise remove a person from their territory, where there are substantial grounds for believing that there is a real risk of irreparable harm, such as that contemplated by articles 6 and 7 of the Covenant, either in the country to which removal is to be effected or in any country to which the person may subsequently be removed.[24]

To sum up, the threshold triggering the non-refoulement principle is a real risk of irreparable harm affecting articles 6 and 7 of the International Covenant on Civil and Political Rights (ICCPR), i.e., the right to life and freedom from inhuman or degrading treatment or punishment of all human beings.

The recent General comment No. 36 (2018) on article 6[25] has also clarified that "Article 6 guarantees this right for all human beings, without distinction of any kind, including for persons suspected or convicted of even the most serious crimes."[26]

It is worth mentioning that non-refoulement principle is, therefore, applicable even to persons suspected or convicted of crimes.[27] Yet, people fleeing environmental disruptions who are outside their countries of origin in vulnerable situations are still excluded from international refugee protection.

This exclusion is even more frustrating because the cited General comment No. 36 (2018) has also acknowledged environmental degradation poses a serious threat to the right to life. It does so in two widely debated paragraphs: paras 26 and 62.

According to para 26,

> The duty to protect life also implies that States parties should take appropriate measures to address the general conditions in society that may give rise to direct threats to life or prevent individuals from enjoying their right to life with dignity. These general conditions may include high levels of criminal and gun violence, [95] pervasive traffic and industrial accidents, [96] degradation of the environment, [97], deprivation of land, territories and resources of indigenous peoples, [98] the prevalence of life threatening diseases, such as AIDS, tuberculosis or malaria, [99] extensive substance abuse, widespread hunger and malnutrition and extreme poverty and homelessness.[28]

Para 62 also reaffirms that

> Environmental degradation, climate change and unsustainable development constitute some of the most pressing and serious threats to the ability of present and future generations to enjoy the right to life. Obligations of States parties under international environmental law should thus inform the contents of article 6 of the Covenant, and the obligation of States parties to respect and ensure the right to life should also inform their relevant obligations under international environmental law. Implementation of the obligation to respect and ensure the right to life, and in particular life with dignity, depends, inter alia, on measures taken by States parties to preserve the environment and protect it against harm, pollution and climate change caused by public and private actors.[29]

In a nutshell, paragraphs 26 and 62 recognize that environmental degradation can be well listed among the "direct" and "most pressing and serious threats" to the ability to enjoy the right to life with dignity. For this reason, States have positive obligations to take "appropriate measures" and "respect and ensure the right to life" by preserving the environment and protecting it against harm, pollution, and climate change caused by public and private actors (Le Moli, 2020).

In light of recent developments, one can conclude that:

a   the term "persecution" can be interpreted as the failure of state protection resulting in systemic human rights violations;

b   situations of generalized violence or events seriously disturbing public order lead to a *prima facie* group-based recognition of refugee status;

c general conditions in society that may give rise to direct threats to life or prevent individuals from enjoying their right to life with dignity also include environmental degradation;

d implementation of States' obligations to ensure the right to life include measures to protect the environment against harm, pollution and climate change caused by public and private actors; and

e the source of "persecution," so conceived, can be the State, a third country, non-state agents, pollution and climate change caused by public and private sectors.

Against this background, why should non-state agents such as polluting industries located in third countries (which tolerate that) not be considered a source of persecution? Why should the effects of nature or human-made disasters not be included among the forms of persecution? Are they not capable of posing serious threats to the environment in which people live and violating their human rights? Why should we ensure international protection to those fleeing wars and violence while excluding those fleeing environmental disruptions?

If the non-refoulement covered those fleeing environmental disruptions as well, there would be numerous advantages. First, those having a well-founded fear of suffering from life-threatening conditions in countries of origin turning uninhabitable would find the international protection they deserve. Second, they would enjoy this protection even if they were from countries that are not a party to any international and regional instruments governing refugee protection.

As noted by UNHCR, the non-refoulement principle enshrined in Article 33 of the 1951 Convention, and complemented by non-refoulement obligations under international human rights law, constitutes a rule of customary international law.[30]

In practice, this means that non-refoulement is a non-derogable jus cogens norm, recognized by the whole international community, that holds States accountable for human rights violations.[31]

The next paragraphs explore how International human rights law and International environmental law may complement and better clarify the scope of such non-refoulement obligations in light of the cited General comment No. 36 (2018) and the historic UN Human Rights Committee's decision in *Teitiota v New Zealand*.[32]

Looking for alternatives: The role of international human rights law and international environmental law

As established by the 1951 Refugee Convention' preamble, the primary aim of the Convention was to assure refugees the widest possible exercise of fundamental rights and freedoms laid down in the Charter of the United Nations and the Universal Declaration of Human Rights approved on 10 December 1948 in the first place. Yet, the UN Charter of 1945, and most notably, the Universal Declaration of Human Rights do not include environmental protection among their main purposes and goals. Nevertheless, environmental protection and human rights are strictly related.

First, they are linked as the environment has an instrumental value for human wellbeing and other recognized human rights such as the right to life, and the right to health.

As enshrined by Article 25(1)

> Everyone has the right to a standard of living adequate for the health and well-being of himself and of his family, including food, clothing, housing and medical care and necessary social services, and the right to security in the event of unemployment, sickness, disability, widowhood, old age or other lack of livelihood in circumstances beyond his control.

Although not legally binding, Stockholm (UN General Assembly, 1972) and Rio (United Nations, 1992) Declarations later clarified how preserving and enhancing the human environment is key to ensuring those fundamental rights.

The first principle enshrined in Stockholm Declaration recalls that

> Man has the fundamental right to freedom, equality and adequate conditions of life, in an environment of a quality that permits a life of dignity and well-being, and he bears a solemn responsibility to protect and improve the environment for present and future generations.[33]

Notably, the 1972 Stockholm Declaration had the undisputed merit of having recognized the links between a healthy environment, human dignity, and human rights for the first time.

Similarly, the second principle laid down by the Rio Declaration also pointed out that

> States have, in accordance with the Charter of the United Nations and the principles of international law, the sovereign right to exploit their own resources pursuant to their own environmental and developmental policies, and the responsibility to ensure that activities within their jurisdiction or control do not cause damage to the environment of other States or of areas beyond the limits of national jurisdiction.[34]

On top of that, a certain level of environmental protection is necessary for protecting the right to life of human beings and preserving the cultural link between individuals, groups, and their environment.[35] A notable trace of this idea is enshrined in Article 27 of the legally binding treaty International Covenant on Civil and Political Rights (ICCPR):

> In those States in which ethnic, religious or linguistic minorities exist, persons belonging to such minorities shall not be denied the right, in community with the other members of their group, to enjoy their own culture, to profess and practice their own religion, or to use their own language.[36]

Not surprisingly, the right to the enjoyment of one's culture is a focus of challenging debates and campaigns advanced by indigenous communities for protecting their territories, natural resources, ancestral lands, languages, beliefs, and ultimately, their cultural survival (Schlosberg & Carruthers, 2010) (see Chapter 4 of this volume). Following the cultural domination, land-grabbing, extractivism by multinational groups, dispossession, and forced displacement they still suffer, their struggles aim to preserve their control over, and relationships with, their environment in order to ensure their "social reproduction":

> the intersecting complex of political-economic, sociocultural, and material-environmental processes required to maintain everyday life and to sustain human cultures and communities on a daily basis and intergenerationally.
>
> (Di Chiro, 2008, p. 281)

Second, the environment also has a value in itself, so that over the last decades, some scholars have observed the emergence of a new human right: the right to a healthy environment (Knox, 2012, 2018). As pointed out by Knox, this right has been incorporated into numerous national constitutions, regional human rights agreements, and treaties since 1972. Examples range from its first integration into the 1976 Portuguese constitution to its recognition by Article 24 of the 1981 African Charter on Human and People's Rights, Article 11 in the 1988 Additional Protocol to the American Convention on Human Rights (Protocol of San Salvador), and Article 1 of the 1988 Arhus Convention. However, although the right to a healthy environment has been increasingly recognized at both domestic and international levels, it has not been explicitly considered to have a human rights nature. A huge gap persists between its legal recognition on paper and its practical implementation through concrete measures to fulfill and promote it (Knox, 2012, 2018).

On top of that, legal recognition of the right to a healthy environment is still in its infancy at the global level. On 8 October 2021, the Human Rights Council (HRC) of the United Nations (UN) recognized the human right to a safe, clean, healthy, and sustainable environment by historic resolution 48/13. Yet, like all HRC resolutions, it is not legally binding.[37] As a result, only citizens of a subset of countries where this right is recognized are entitled to enjoy a healthy environment. Accordingly, this right is still conceived as a partially recognized right.

At the time of writing, the instrumental value of the environment to ensure other fundamental rights such as the right to life, health, food, water, housing, culture, development, etc., seems to be the most viable way to bind states non-refoulement obligations under international human rights. As it has been stressed by international human rights bodies, environmental degradation can seriously affect those fundamental rights, thus compromising the possibility to return people to places turning uninhabitable. In principle, the non-refoulement can be used to prevent states from returning asylum seekers to a country where environmental degradation poses serious threats to life (Scott, 2014).

In practice, however, the issue is far from straightforward due to the extremely high threshold required to establish a real risk of irreparable harm.[38]

In this regard, main goals and core principles of International Environmental Law (IEL) are particularly significant in supporting this argument for various reasons.

The first relies on environmental protection as a crucial goal of IEL. Its primary aim is to control pollution and depletion of natural resources without compromising economic development. Such a challenging balance between environmental protection and economic development has been faced by introducing the contested notion of sustainable development. This term was formally introduced by the 1987 Report of the World Commission on Environment and Development: Our Common Future (World Commission on Environment and Development, 1987). Known as the Brundtland Report, it enshrined that sustainable development implies confining economic development by ecological limits in order to meet "the needs of the present without compromising the ability of future generations to meet their own needs" (World Commission on Environment and Development, 1987, para. 27). This concept was also included in the 1992 Rio Declaration that recognized environmental protection as a necessary precondition to achieve sustainable development and then equally fulfill the environmental needs of present and future generations.

Over the years, policies advanced to cope with this ambitious goal have, however, failed to control pollution and mitigate the effects of climate change. Since 1990, IPCC reports have provided clear evidence that human activities play an increasingly important role in global warming.

In particular, the 2014 Fifth Assessment Report (AR5) provided the scientific input into the Paris Agreement by extensively recognizing that greenhouse gas emissions are likely to cause further warming, long-lasting changes in the climate system, and irreversible impacts on people and ecosystems (IPCC, 2014). With this report, IPCC clearly stated that it was "95 percent certain that humans are the main cause of current global warming" (IPCC, 2014, p. v).[39]

Notably, in 2015 States signed the Paris Agreement, a legally binding treaty that poses obligations to states to keep the temperature below 2 degrees. The main assumption was that developed states had contributed most to global warming. Therefore, they have common but differentiated responsibilities urging them to take the lead in undertaking emission reduction targets, providing financial resources, and enhancing support for capacity-building actions in developing countries Parties. In compliance with the Paris Agreement, States have positive duties to preserve environment, and to protect it against harm, pollution, and climate change. As reaffirmed in the General comment No. 36 (2018), these obligations under international environmental law should inform the contents of article 6 of the Covenant to protect life from serious threats arising from environmental degradation.

Over the years, IEL has also established useful legal principles to protect the environment while improving the overall environmental quality by containing pollution and environmental changes.[40] The most relevant for showing how IEL may contribute to trigger non-refoulement obligations for people fleeing environmental disruptions are precautionary principle, polluter pays principle, common but differentiated responsibility and respective capabilities.[41]

The precautionary principle derives from the German *Vorsorgeprinzip*, conceived as an action principle combining "caution with caring for the future, as well as providing for it" (O'Riordan, Cameron, & Jordan, 2001, p. 16). It evolved out of the 1970s lively debate that introduced this principle under the West German environmental law (Von Moltke, 1987).

Usually explained by the formula "better safe than sorry," the precautionary principle advocates states to take anticipatory action to protect the environment while preventing environmental damages despite the lack of scientific certainty.

As for the international context, the 1992 Rio Declaration enshrined the entry of the precautionary principle among IEL's key principles by the following Principle 15:

> In order to protect the environment, the precautionary approach shall be widely applied by States according to their capabilities. Where there are threats of serious or irreversible damage, lack of full scientific certainty shall not be used as a reason for postponing cost-effective measures to prevent environmental degradation.[42]

Even though one may object that the 1992 Rio Declaration is a legally non-binding instrument, in the United Nations Framework Convention on Climate Change (UNFCCC), which is legally binding, the precautionary principle has been subsequently reformulated as follows:

> The Parties should take precautionary measures to anticipate, prevent or minimize the causes of climate change and mitigate its adverse effects. Where there are threats of serious or irreversible damage, lack of full scientific certainty should not be used as a reason for postponing such measures, taking into account that policies and measures to deal with climate change should be cost-effective so as to ensure global benefits at the lowest possible cost.[43]

Following this formulation, one can observe that states have positive duties to take anticipatory measures, even in the absence of scientific certainty, if the threshold of serious or irreversible damage is proved. In practical terms, the main advantage of applying the precautionary principle for granting protection to the "climate refugees" is that it makes it not necessary to establish the causal link between climate change and migration/displacement as scientific evidence is neither required nor can be used for not taking action (Poon, 2018).

As pointed out by Poon, however, the precautionary principle has to be used in conjunction with non-refoulement for protection being granted to climate refugees. The (sole) precautionary principle cannot address the harm caused to "climate refugees," as its main focus is on irreversible damages to the environment within the confines of one State's domestic jurisdiction.

Similarly, despite its extraterritorial application, the (sole) non-refoulement does not protect those fleeing environmental disruptions, as they are not listed on the grounds of persecution.

Thus, in Poon's view, protection to "climate refugees" can be ensured only by using those principles together to extend the extraterritorial application of the non-refoulement to the precautionary principle.

Other scholars have also pursued Poon's call for bridging international environmental law and international refugee law (Ahmed, 2018; Eckersley, 2015; Kent & Behrman, 2018). Among them, Ahmed has tried to trigger non-refoulement obligations by employing the polluter pays principle. This principle first appeared in a 1972 Council Recommendation on Guiding Principles Concerning the International Economic Aspects of Environmental Policies of the Organisation for the Economic Co-operation and Development (OECD). The recommendation stated that:

> the polluter should bear the expenses of carrying out the above-mentioned measures decided by public authorities to ensure that the environment is in an acceptable state. In other words, the cost of these measures should be reflected in the cost of goods and services which cause pollution in production and/or consumption. Such measures should not be accompanied by subsidies that would create significant distortions in international trade and investment.[44]

The polluter pays principle was then reaffirmed in various documents[45] until it was formally recognized as a fundamental principle of IEL by Principle 16 of the Rio Declaration. It laid down that the "polluter should, in principle, bear the cost of pollution, with due regard to the public interest and without distorting international trade and investment."[46]

In light of this principle, Ahmed proposes a model for resettling "climate refugees" in those liable countries that have contributed most to Greenhouse gas emissions and pollution. In doing so, the author designs a climate refugee settlement model based on four parameters: per capita CO_2 emissions (2011), per capita GNI (2015), human development index (HDI) 2014, and per capita planet's resource consumption (Ahmed, 2018). The author also proposes an equation to establish which liable countries are expected to take responsibility for resettling "climate refugees," and by what proportion. The main assumption is that "liable countries should pay for polluting the climate and share the burden proportionally" (Ahmed, 2018, p. 21). While this assumption certainly raises important environmental justice issues, it does not seem to overcome the challenges of demonstrating the causal link between climate change and human activities. More specifically, the polluter pays principle does not adequately address climate-related harms that cannot be attributed to specific polluters (Eckersley, 2015). Given the unwillingness of States to implement this principle effectively, a more promising pathway seems to use it in the field of climate change litigation (Eckersley, 2015; Kent & Behrman, 2022).

Ahmed's model, however, has had the merit to take into due account greater past emissions generated by developed countries, thus leading us to examine the applicability of the common but differentiated responsibility and respective capabilities to trigger non-refoulement obligations.

The common but differentiated responsibility and respective capabilities is one of the pivotal IEL principles. Initially established by Article 23 of Stockholm Declaration, it was reaffirmed by Principle 7 of Rio Declaration, until it has been acknowledged among the pillars of the climate regime by Article 3(1) of the 1992 UNFCCC. This latter defined the principle as follows:

> The Parties should protect the climate system for the benefit of present and future generations of humankind, on the basis of equity and in accordance with their common but differentiated responsibilities and respective capabilities. Accordingly, the developed country Parties should take the lead in combating climate change and the adverse effects thereof.[47]

The novelty introduced by this principle is that obligations on States should vary depending upon their economic development, circumstances, and capabilities. In other words, not only developed countries contributed most to global warming, but they also have more resources and capabilities to combat the effects of climate change. In this regard, the suitability of this principle to trigger non-refoulement obligations sounds promising as it may better address disproportionate responsibilities for having caused climate change and financial and technological capabilities to tackle its effects.

In this regard, its applicability to the "climate refugees' case has been extensively tested by Robyn Eckersley (Eckersley, 2015). According to the author, the search for recognition of "climate refugees" leads us to rethink the "grammar" of responsibility in the first place. In light of the complex interdependence and unclear chain of causation, a liability model will be rarely successful in identifying guilty perpetrators. Nor it will capture the main "culprit:" the "social structures that enable the systematic generation of harm" (Eckersley, 2015; Young, 2006).

According to Eckersley, complex interdependence has to be read in conjunction with feasible solutions. In her view, one should consider that the most capable might not be the most interested or motivated countries to take action in favor of the most vulnerable countries.

Accordingly, she suggests distinguishing from States responsibilities to provide financial assistance to "climate refugees" to their responsibilities to receive them.

Such a distinction may prevent States from resisting to recognize "climate refugees" right to choose their host countries as a form of reparation for past injustice, loss, and damage.

Also, it may "enable receiving states and climate refugees to receive financial and other assistance to defray the costs of resettlement"(Eckersley, 2015, p. 495). In practical terms, the more UNFCCC parties can offer compensation to "climate refugees" by the Warsaw International Mechanism, the more States are likely to take responsibility to receive and accept "climate refugees" right to choose their host communities.

To sum up, Eckersley argues that this solution would make ethically desirable and politically feasible granting the right of "climate refugees" to choose their host states at present time. In the longer term, instead, she reaffirms the need for a hybrid approach combining different principles for allocating responsibility for "climate refugees."

Any model of responsibility elaborated in the longer term should bridge the mismatch between culpable actors who have caused and benefitted from pollution and resource depletion, and capable actors who can remedy injustices of the most affected or prevent injustices from being perpetrated.

In this view, she suggests creating a global insurance system or Climate Super-fund as a concrete measure to make "culpable actors" contribute while most vulnerable draw (Eckersley, 2015).

To conclude, human rights obligations related to the enjoyment of a healthy environment on the one hand, and international environmental law obligations to protect the environment, on the other hand, suggest that states must take all necessary measures to address the general conditions of society able to generate indirect and direct threats to the right to life. Those include environmental degradation and pollution, even if those have not resulted in the loss of life (Le Moli, 2020). On top of that, IEL principles read in conjunction with human rights and or international refugee law concepts significantly contribute to triggering non-refoulement obligations.

Although in theory, scholars' proposals advance promising future perspectives while showing the potential of exploring strategies beyond silos and the boundaries of various fields of law. In practice, some of these proposals and reflections have been tested in a landmark decision of the UN Human Rights Committee (HRC) –Teitiota v. New Zealand, relating to climate change as a threat to life. The following session examines this decision by a critical stand.

Ioane Teitiota v. New Zealand – a legal tipping point?

The case Teitiota v. New Zealand has offered the possibility to investigate if and to what extent climate change may be conceived as a threat to life able to trigger non-refoulement obligations under international human rights law.

The case concerns Ioane Teitiota, a national of the Republic of Kiribati. In 2013, Teitiota claimed that he and his family were forced to migrate from the island of Tarawa in the Republic of Kiribati to New Zealand due to sea level rise caused by global warming. Saltwater contamination, lack of fresh water supply, housing crises, land disputes, overcrowding were the climate change impacts mentioned by Teitiota, which made Kiribati an untenable and violent environment for him and his family. On this ground, he applied for refugee status in New Zealand in 2013.

The Immigration and Protection Tribunal rejected his claim for asylum but did not exclude the possibility that environmental degradation could "create pathways into the Refugee Convention or protected person jurisdiction." Teitiota's subsequent appeals to the Court of Appeal (2014) and the Supreme Court (2015) were also unsuccessful. Both Courts confirmed that there was no evidence establishing substantial grounds for a violation of Teitiota and his family's rights under Article 6 of the Covenant by returning to Kiribati.

Thus, on 15 September 2015, Teitiota received a deportation order. On 23 September 2015, he was then removed to Kiribati, and his family left shortly thereafter. Teitiota argued that the assessment of the domestic authorities amounted to a denial of justice, and filed a complaint before the Human Right Committee (HRC). In particular, he claimed that his removal to the Republic of Kiribati violated his rights under article 6 (1) of the Covenant.

For the sole purpose of admissibility, the HRC declared that his complaint was supported by sufficient evidence that adverse effects of climate change correlated with sea level rise he faced upon deportation to the Republic of Kiribati resulted in a real risk of impairment to his right to life under article 6 of the Covenant (para 8.6). In the consideration on merits, the HRC made clear it had to ascertain whether a real risk of irreparable harm to Teitiota's life occurred and if there was a denial of justice in the assessment by the State party's authorities when he was removed to the Republic of Kiribati.

Its decision was based on two general comments: general comment No. 31 (2004) on the nature of the general legal obligation imposed on States parties to the Covenant, and general comment No. 36 (2018) on article 6 of the Covenant on the right to life.

By recalling its general comments, the HRC clarified that one should have demonstrated that (1) there were substantial grounds for believing that there was a real risk of irreparable harm such as that contemplated by articles 6 and 7 of the Covenant (general comment No. 31, para 12) and (2) the State party's acts or omissions resulted in a violation of the right to life, or posed an existing or imminent threat to the enjoyment of such right (general comment No. 36, para 7).

Obligations not to deport pursuant article 6 of the Covenant have a broader scope than the non-refoulement under the 1951 Refugee Convention. Those also include people who are not entitled to refugee status but nonetheless deserve protection against refoulement (general comment No. 36, para 31).

Although Teitiota may fall under this broader interpretation, a major obstacle is represented by the real risk threshold. As noted by HRC, the threshold is high and has not been met in this case. The difficulty of meeting the threshold emerges from the following aspects.

As for the risk of suffering serious physical harm from violence linked to land disputes and overcrowding mentioned by Teitiota, HRC concludes that the threshold has not been met. Sporadic incidents of violence between land claimants are not enough to demonstrate the real, personal and reasonably foreseeable risk of irreparable harm specific to him. Those episodes have not even involved Teitiota and were faced by all individuals in Kiribati.

The high risk threshold has not also been met in the water and food security-issues raised by Teitiota. Providing evidence that 60 percent of residents of South Tarawa get fresh water from rationed supplies does not demonstrate the lack of access to potable water. For the high risk threshold to be met, Teitiota should have demonstrated that the fresh water supply was *inaccessible, insufficient or unsafe.*

Along the same line of reasoning, he also failed to demonstrate the deprivation of means of subsistence. Even if salt deposits on the ground make it difficult to grow crops, they do not make it impossible. HRC notes that most nutritious crops remained available in the Republic of Kiribati. Further, alternative sources of employment or financial assistance might have helped Teitiota meeting his basic needs. As a result, HCR concludes that Teitiota failed to demonstrate that if deported to Kiribati he would suffer *indigence, deprivation of food, and extreme precarity*.

A further obstacle preventing Teitiota from meeting the high-risk threshold has to do with author's claim that climate change is rendering the Republic of Kiribati uninhabitable. The timeframe of 10 to 15 years estimated for the Republic of Kiribati to become uninhabitable has been considered by HCR as a reasonable time for the government, with the assistance of the international community, to prevent the island's submersion and, where necessary, to relocate its population. This consideration leads us to the general comment No. 36 (2018) on article 6 of the Covenant on the right to life also recalled by HCR. General comment No. 36 (2018) establishes that the protection of the right to life requires States parties to adopt positive measures. As enshrined by para 21,

> States parties are thus under a due diligence obligation to undertake reasonable positive measures, which do not impose on them disproportionate burdens, in response to reasonably foreseeable threats to life originating from private persons and entities, whose conduct is not attributable to the State.[48]

Although the general comment No. 36 uses the wording "reasonably foreseeable threats to life," HCR has focused more on impacts already felt in Kiribati to date rather than on threats to life associated with climate change which are reasonably foreseeable today, but may occur in the next decade or so. In doing so, it has also undermined the close link between international human rights law and international environmental law established by para 62 as follows:

> Environmental degradation, climate change and unsustainable development constitute some of the most pressing and serious threats to the ability of present and future generations to enjoy the right to life. Obligations of States parties under international environmental law should thus inform the contents of article 6 of the Covenant, and the obligation of States parties to respect and ensure the right to life should also inform their relevant obligations under international environmental law.[49]

In other words, the narrow interpretation of HCR may result in a *probatio diabolica*, given

> the limited space left to international environmental law to inform the contents of article 6 of the Covenant. Indeed, HCR has put an unreasonable burden of proof on the author to establish the real risk of a violation of his right

to life by setting an almost unreachable high-risk standard.[50] By contrast, it has not taken into due account State Party's compliance with its positive duty to protect Teitiota's life from risks arising from sea level rise. (e.g., scarcity of potable water)

As observed by one of the dissenting experts on the committee, Vasilka Sancin, HCR assessed that the Government of Kiribati had taken adequate measures to address impacts of climate change by adopting the 2007 National Adaptation Programme of Action. However, the 2008 National Water Resources Policy and a 2010 National Sanitation Policy's priorities have not been yet implemented.[51]

If more space was left to international law to inform the contents of article 6 of the Covenant, much more emphasis would be devoted on the duty of State's Party to protect its citizens from environmental harm, thus recognizing the interlinkages between human rights and international environmental law.

As recalled by McAdam, HRC might have also considered international environmental law principles, such as prevention, international cooperation, precautionary, and intergenerational justice principles. As she pointed out, such principles would have helped to

(i)   prevent foreseeable domestic and extraterritorial human rights violations resulting from climate change;
(ii) cooperate internationally in the face of the global climate emergency;
(iii) apply the precautionary principle to protect life in the face of uncertainty; and
(iv) ensure intergenerational justice for children and posterity. (McAdam, 2020, p. 717)

McAdam also underlined that HRC somehow neglected the potential future harm however included in the refugee definition. In her view, HRC extensively focused on the certainty of harm while disregarding the existence of a real risk of it. In doing so, HRC relied significantly on adaptation plans or potential mitigation developments eventually enacted by Kiribati State, but those might not be sufficient to tackle an existing real risk (McAdam, 2020). In light of this, she suggested replacing the "imminence" of irreparable harm with the "likelihood of harm" as a criterium to establish the lawfulness of repatriation. Such a replacement would avoid that HRC implicitly advanced future-conditional protection for present harms. Indeed, Kiribati is *already* facing the effects of climate change. As pointed out by Behrman and Kent, while Courts wait for environmental conditions to become sufficiently worse to trigger non-refoulement obligations, increasing numbers of people like Teitiota are legally trapped with the "present" effects of climate change (Behrman & Kent, 2020). On top of that, HRC set a stringent high-risk standard to establish that a real risk of irreparable harm exists, but it did not establish an equally stringent standard to establish States' compliance with their existing positive duties to protect the environment and human rights depending on it.

In light of these considerations, is this historic HRC decision a legal tipping point or still conservative? Did it help understand to what extent climate change may be conceived as a threat to life able to trigger non-refoulement obligations?

As for the first question, the HRC decision has been innovative in accepting and recognizing the potential of climate change to trigger non-refoulement obligations on principle. Although the underlying legal principles were already established by General Comment No. 36, the formal recognition of their potential use concerning non-refoulement can make a difference for future law cases (cf. McAdam, 2020). The HRC has been, however, conservative in its practical application. As explained earlier, the stringent high-risk threshold required and the double standard applied to States' positive duties laid down in international environmental law make its application challenging and potentially unfair.

As for the second question, the HRC has certainly met the goal of clarifying in-depth the necessary requirements to trigger non-refoulement obligations. For example, HRC explained that sporadic incidents of violence do not succeed in proving the real, personal, and reasonably foreseeable risk of irreparable harm. Indeed, this latter should be specifically directed to the plaintiff (e.g., Teitiota). A further clarification concerns the threshold for raising water-security issues effectively. The threshold being met, is not sufficient to prove that freshwater can only be obtained from rationed supplies. Rather, the plaintiff should demonstrate that the freshwater supply is *inaccessible, insufficient, or unsafe.* A similar argument has also been advanced for food-security issues. According to HRC, food scarcity is not enough to demonstrate the exposure to *indigence, deprivation of food, and extreme precarity* if most nutritious crops remain available and alternative sources of employment and financial assistance are offered by the home country. Finally, HRC also clarified when uninhabitability can be "declared": when the timeframe left does not allow for intervening acts by the sending State.

To conclude, climate refugees' recognition still has a long way to go, but this HRC decision represents a meaningful step forward if only to understand the necessary and sufficient conditions better to trigger non-refoulement obligations.

Notes

1 Hereinafter the 1951 Convention.
2 Gilbert Jaeger reports that people leaving Russian territories for countries of Europe or Asia minor amounted to estimates varying between 1 million and 2 million (Jaeger, 2001).
3 By this expression, the Convention meant Assyrians, Assyro-Chaldeans, Syrians, Kurds and a small number of Turks (cf. Jaeger, 2001).
4 See UN Ad Hoc Committee on Refugees and Stateless Persons, A Study of Statelessness, United Nations, August 1949, Lake Success – New York, 1 August 1949, E/1112; E/1112/Add.1, available at: www.refworld.org/docid/3ae68c2d0.html [accessed 9 November 2021].
5 See UN Relief and Works Agency for Palestine Refugees in the Near East (UNRWA), Consolidated Eligibility and Registration Instructions (CERI), 1 January 2009, available at: www.refworld.org/docid/520cc3634.html [accessed 9 November 2021].
6 Cf. recommendation 773 (1976) *on the Situation of De Facto Refugees* by the Council of Europe, defining de facto refugees as "all persons not recognised as refugees within the meaning of Article 1 of the Convention relating to the Status of Refugees of 28 July 1951 as amended by the Protocol of 31 January 1967."

7 Hurwitz highlighted this state-centric approach noting that: "Since the first Nansen arrangement of 1922, as in the interwar arrangements, the essential condition of refugees is their presence outside the territory of their country and their lack of protection by any State. However, the Convention added to these criteria the condition of a well-founded fear of persecution based on one or more of five grounds: race, religion, nationality, political opinion, and membership of a social group. Refugee status determination is carried out by State parties which are free to institute procedures they see appropriate for this purpose" (Hurwitz, 2009, p. 13).

8 The authors claim to incorporate a category of "environmental persecution" within the 1951 Refugee Convention, thus expanding its scope and recognizing climate refugees as a new category of refugees.

9 Guiding Principles on Internal Displacement, UN Doc E/CN.4/1998/53/Add.2 (adopted 11 February 1998).

10 African Union Convention for the Protection and Assistance of Internally Displaced Persons In Africa (adopted 22 October 2009, entered into force 6 December 2012), Article 5, para 4.

11 See the comment of Mona Rishmawi, an Independent Expert of the UN Human Rights Commission for Somalia, given during the hearing of 29 November 1999 before the Committee for Human Rights and Humanitarian Aid of the German Parliament (European Council on Refugees and Exiles, 2000).

12 See Organization of African Unity (OAU), Convention Governing the Specific Aspects of Refugee Problems in Africa ("OAU Convention"), 10 September 1969, 1001 U.N.T.S. 45, available at: www.refworld.org/docid/3ae6b36018.html [accessed 9 November 2021].

13 For a comprehensive overview of the 1969 OAU Convention's influence outside Africa, see (Sharpe, 2018).

14 UN High Commissioner for Refugees (UNHCR), *Procedural Standards for Refugee Status Determination Under UNHCR's Mandate*, 26 August 2020, available at: www.refworld.org/docid/5e870b254.html [accessed 2 November 2020].

15 In this regard, see also President Biden's Executive Order 14013 (4 February 2021) on "Rebuilding and Enhancing Programs to Resettle Refugees and Planning for the Impact of Climate Change on Migration." Following this Executive Order, a Task Force has been designed to inform the report requested by President Biden in Section 6 (Climate Change and Migration). The Task Force publication is available at www.refugeesinternational.org/reports/2021/7/12/task-force-report-to-the-president-on-the-climate-crisis-and-global-migration-a-pathway-to-protection-for-people-on-the-move [accessed 10 November 2021].

16 In 2019, coal supplied 58% of China's total energy consumption (Cf U.S. Energy Information Administration, 2020).

17 At this early stage of development, it is too soon to tell whether their commitments are confirmed, as most will depend on the outcome of the COP26 in Glasgow.

18 Cf. Global Compact for Safe, Orderly and Regular Migration, para 15.

19 In this regard, see Objective 2 in Global Compact for Safe, Orderly and Regular Migration.

20 See Objective 23 in Global Compact for Safe, Orderly and Regular Migration.

21 See Global Compact for refugees, paras 78–79.

22 See UN General Assembly, Convention Relating to the Status of Refugees, 28 July 1951, United Nations, Treaty Series, vol. 189, Article 1A, available at: www.refworld.org/docid/3be01b964.html [accessed 19 June 2021].

23 UN High Commissioner for Refugees (UNHCR), Note on International Protection, 12 July 2006, A/AC.96/1024, available at: www.refworld.org/docid/44c9cb212.html [accessed 8 February 2021], para 5.

24 UN Human Rights Committee (HRC), General comment no. 31 [80], The nature of the general legal obligation imposed on States Parties to the Covenant, 26 May 2004,

CCPR/C/21/Rev.1/Add.13, available at: www.refworld.org/docid/478b26ae2.html [accessed 10 February 2021]

25 UN Human Rights Committee (HRC), General comment no. 36, Article 6 (Right to Life), 3 September 2019, CCPR/C/GC/35, available at: www.refworld.org/docid/5e5e75e04. html [accessed 16 February 2021].

26 Ibid. para 3.

27 Cf. CJEU Joined Cases C-391/16, C-77/17 and C-78/17 / Judgment M v Ministerstvo Vnitra (C-391/16) and X (C-77/17), X (C-78/17) v Commissaire général aux réfugiés et aux apatrides, available at: http://curia.europa.eu/juris/document/document_print. jsf?docid=214042&text=&dir=&doclang=EN&part=1&occ=first&mode=lst&pageIn dex=0&cid=5124920 [accessed 17 February 2021].

28 UN Human Rights Committee (HRC), General comment no. 36, Article 6 (Right to Life), 3 September 2019, CCPR/C/GC/35, available at: www.refworld.org/docid/5e5e75e04. html [accessed 16 February 2021], para 26.

29 Ibid. para 62.

30 Cf. UN High Commissioner for Refugees (UNHCR), Advisory Opinion on the Extra-territorial Application of Non-Refoulement Obligations under the 1951 Convention relating to the Status of Refugees and its 1967 Protocol, 26 January 2007, available at: www.refworld.org/docid/45f17a1a4.html [accessed 18 February 2021]; see also UN High Commissioner for Refugees (UNHCR), The Principle of Non-Refoulement as a Norm of Customary International Law. Response to the Questions Posed to UNHCR by the Federal Constitutional Court of the Federal Republic of Germany in Cases 2 BvR 1938/93, 2 BvR 1953/93, 2 BvR 1954/93, 31 January 1994, available at: www. refworld.org/docid/437b6db64.html [accessed 17 February 2021].

31 For an overview of how States have applied such non-refoulement obligations, see also (Newmark, 1993).

32 UN Human Rights Committee (HRC), *Ioane Teitiota v New Zealand* UN Doc CCPR/ C/127/D/2728/2016 (7 January 2020).

33 UN General Assembly, United Nations Conference on the Human Environment, 15 December 1972, A/RES/2994, available at: www.refworld.org/docid/3b00f1c840.html [accessed 17 September 2021].

34 A/CONF.151/26 (Vol. I) Report of the United Nations Conference on Environment and Development, available at: www.un.org/en/development/desa/population/migra-tion/generalassembly/docs/globalcompact/A_CONF.151_26_Vol.I_Declaration.pdf [accessed 17 September 2021].

35 For a detailed examination of the potential use of the right to enjoy culture their own culture to protect climate migrants, see (Wewerinke-Singh, 2018).

36 UN General Assembly, International Covenant on Civil and Political Rights, 16 December 1966, United Nations, Treaty Series, vol. 999, p. 171, available at: www.ref-world.org/docid/3ae6b3aa0.html [accessed 17 September 2021]. See also UN Human Rights Committee (HRC), CCPR General Comment No. 23: Article 27 (Rights of Minorities), 8 April 1994, CCPR/C/21/Rev.1/Add.5, available at: www.refworld.org/ docid/453883fc0.html [accessed 17 September 2021].

37 See resolution A/HRC/48/L.23/Rev.1 on the Human right to a safe, clean, healthy and sustainable environment available at https://documents-dds-ny.un.org/doc/UNDOC/ LTD/G21/270/15/PDF/G2127015.pdf?OpenElement [accessed 10 November 2021].

38 See UN Human Rights Committee (HRC), *Ioane Teitiota v New Zealand* UN Doc CCPR/ C/127/D/2728/2016 (7 January 2020) as analyzed in the following section of this volume.

39 The IPCC has reaffirmed this position by its 2021 report (HYPERLINK "15032-5135-Full-Book.docx" \l "Ref_106_FILE150325135002" \o "(AutoLink):IPCC. (2021). IPCC, 2021: Summary for Policymakers. In R. Y. and B. Z. Masson Delmotte, V., P. Zhai, A. Pirani, S. L. Connors, C. Péan, S. Berger, N. Caud, Y. Chen, L. Goldfarb, M. I. Gomis, M. Huang, K. Leitzell, E. Lonnoy, J. B. R. Matthews, T. K. Maycock, T. Waterfield, O. Yelekçi (Ed.), Climate Change 2021: The Physical Science Basis. Contribution of Working Group I to the

Sixth Assessment Report of the Intergovernmental Panel on Climate Change (pp. 1–41). Cambridge University Press. UserName - DateTime: mbl-12/17/2021 4:37:39 PM" IPCC, 2021), warning that unprecedented human-induced climate changes are already affecting many weather and climate extremes across the globe. These latter have irreversible effects in many cases and urged the promotion of adaptation and mitigation policies to prevent, among others, large-scale displacements.

40 For a detailed overview, see (Kravchenko, Chowdhury, & Law, 2012).
41 For a comprehensive overview of all environmental law principles' contribution to coping with "climate refugees' issue, see (Kent & Behrman, 2018, Chapter 3).
42 Rio Declaration on Environment and Development (adopted 14 June 1992) UN Doc A/CONF.151/26 (vol. I)/31 ILM 874 (1992) principle 15.
43 See United Nations Framework Convention on Climate Change (UNFCCC), 20 January 1994, UN Doc A/RES/48/189, art 3(3).
44 OECD, *Council Recommendation on Guiding Principles Concerning the International Economic Aspects of Environmental Policies of the Organization for Economic Co-operation and Development* (1972), C (72) 128, para 4.
45 For a detailed overview, see (Kravchenko et al., 2012).
46 Rio Declaration on Environment and Development (adopted 14 June 1992) UN Doc A/CONF.151/26 (vol. I)/31 ILM 874 (1992) principle 16.
47 See United Nations Framework Convention on Climate Change (UNFCCC), 20 January 1994, UN Doc A/RES/48/189, art 3(1).
48 See UN Human Rights Committee (HRC), General comment no. 36, Article 6 (Right to Life), 3 September 2019, CCPR/C/GC/35, available at: www.refworld.org/docid/5e5e75e04.html [accessed 16 February 2021].
49 Ibid.
50 See the Individual opinion of Committee member Duncan Laki Muhumuza in UN Human Rights Committee (HRC), *Ioane Teitiota v New Zealand* UN Doc CCPR/C/127/D/2728/2016 (7 January 2020), Annex 2.
51 See the Individual opinion of Committee member Vasilka Sancin in UN Human Rights Committee (HRC), *Ioane Teitiota v New Zealand* UN Doc CCPR/C/127/D/2728/2016 (7 January 2020), Annex 1.

Reference list

Abebe, T. T., Abebe, A., & Sharpe, M. (2019). *The 1969 OAU refugee convention at 50. ISS Africa report*. Geneva: UNHCR. https://media.africaportal.org/documents/the_1969_OAU_Refugee_convention.pdf

Ahmed, B. (2018). Who takes responsibility for the climate refugees? *International Journal of Climate Change Strategies and Management, 10*(1), 5–26. https://doi.org/10.1108/IJCCSM-10-2016-0149

Andrade, J. H. F. De. (2008). On the development of the concept of persecution in international refugee law. *Anuário Brasileiro De Direito Internacional |Brazilian Yearbook of International Law, 2*(3), 114–136.

Appleby, J. K., & Kerwin, D. (2018). *International migration policy report: Perspectives on the content and implementation of the global compact for safe, orderly, and regular migration* (Vol. 73). New York. https://doi.org/10.14240/internationalmigrationrpt2018

Behrman, S., & Kent, A. (2020). The teitiota case and the limitations of the human rights framework. *Questions of International Law, 75*, 25–39.

Betts, A. (2016). *Survival migration. Survival migration*. Cornell University Press. https://doi.org/10.7591/cornell/9780801451065.003.0002

Bufalini, A. (2019, April). The global compact for safe, orderly and regular migration: What is its contribution to international migration law? *QIL, 4*, 5–24.

Cantor, D. J. (2019). Fairness, failure, and future in the refugee regime. *International Journal of Refugee Law, 30*(4), 627–629. https://doi.org/10.1093/ijrl/eey069

Coles, G. J. L. (1990). *Placing the refugee issue on the new international agenda*. Geneva: United Nations High Commission for Refugees.

Conisbee, M., & Simms, A. (2003). *Environmental refugees: The for recognition*. London: New Economics Foundation.

Cooper, J. B. (1997). Environmental refugees: Meeting the requirements of the refugee definition. *New York University Environmental Law Journal, 6*(2), 480–529.

Di Chiro, G. (2008). Living environmentalisms: Coalition politics, social reproduction, and environmental justice. *Environmental Politics, 17*(2), 276–298. https://doi.org/10.1080/09644010801936230

Docherty, B., & Giannini, T. (2009). Confronting a rising tide: A proposal for a convention on climate change refugees. *Harvard Environmental Law Review, 33*(2), 349–403.

Eckersley, R. (2015). The common but differentiated responsibilities of states to assist and receive 'climate refugees.' *European Journal of Political Theory, 14*(4), 481–500. https://doi.org/10.1177/1474885115584830

El-Hinnawi, E. (1985). *Environmental refugees*. Nairobi: United Nations Environment Programme.

European Council on Refugees and Exiles. (2000). *Non-state agents of persecution and the inability of the state to protect: The German interpretation*. London: ECRE.

Fischel De Andrade, J. H. (2019). The 1984 Cartagena declaration: A critical review of some aspects of its emergence and relevance. *Refugee Survey Quarterly, 38*(4), 341–362. https://doi.org/10.1093/rsq/hdz012

Foster, M. (2009). Non-refoulement on the basis of socio-economic deprivation: The scope of complementary protection in international human rights law. *New Zealand Law Review*, (2), 257–310.

Gemenne, F. (2017). The refugees of the anthropocene. In B. Mayer & F. Crépeau (Eds.), *Research handbook on climate change, migration and the law* (pp. 394–404). https://doi.org/10.4337/9781785366598.00025

Goodwin-Gill, G. S., & McAdam, J. (2007). *The refugee in international law*. Oxford: Oxford University Press.

Hathaway, J. C. (2017). *Reconceiving refugee law as human rights protection, Refugees and Rights* (M. Crock, Ed.). Osgoode Hall Law School, York University: Taylor and Francis. https://doi.org/10.4324/9781315244969

Hurwitz, A. (2009). *The collective responsibility of states to protect refugees. The collective responsibility of states to protect refugees*. New York: Oxford University Press. https://doi.org/10.1093/acprof:oso/9780199278381.001.0001

IDMC. (2021). *Global report on internal displacement*. Geneva: IDMC.

Ionesco, D. (2017). *The atlas of environmental migration*. Routledge. https://doi.org/10.4324/9781315777313

IPCC. (2014). *Climate change 2014. Synthesis report*. Geneva: IPCC.

IPCC. (2021). IPCC, 2021: Summary for policymakers. In B. Z. Masson Delmotte, V., P. Zhai, A. Pirani, S. L. Connors, C. Péan, S. Berger, N. Caud, Y. Chen, L. Goldfarb, M. I. Gomis, M. Huang, K. Leitzell, E. Lonnoy, J. B. R. Matthews, T. K. Maycock, T. Waterfield, & O. Yelekçi (Eds.), *Climate change 2021: The physical science basis. contribution of working group I to the sixth assessment report of the intergovernmental panel on climate change* (pp. 1–41). Cambridge University Press.

Jackson, I. C. (1999). *The refugee concept in group situations*. The Hague and London: Martinus Nijhoff Publishers.

Jaeger, G. (2001). On the history of the international protection of refugees. *Revue Internationale de La Croix-Rouge/International Review of the Red Cross*. https://doi.org/10.1017/s1560775500119285

Kelley, C. P., Mohtadi, S., Cane, M. A., Seager, R., & Kushnir, Y. (2015). Climate change in the fertile crescent and implications of the recent Syrian drought. *Proceedings of the National Academy of Sciences, 112*(11), 3241–3246. https://doi.org/10.1073/pnas.1421533112

Kent, A., & Behrman, S. (2018). *Facilitating the resettlement and rights of climate refugees: An argument for developing existing principles and practices*. Abingdon: Routledge. https://doi.org/10.4324/9781351175708

Kent, A., & Behrman, S. (Eds.). (2022). *Climate refugees: Global, local and critical approaches*. Cambridge: Cambridge University Press. https://doi.org/DOI:

Knox, J. H. (2012). *Report of the independent expert on the issue of human rights obligations relating to the enjoyment of a safe, clean, healthy and sustainable environment, preliminary report*. Geneva: United Nations Human Rights Council. https://www.ohchr.org/Documents/HRBodies/HRCouncil/RegularSession/Session22/A-HRC-22-43_en.pdf

Knox, J. H. (2018). *Human rights obligations relating to the enjoyment of a safe, clean, healthy and sustainable environment. Note by the secretary-general*. New York: United Nations General Assembly.

Kravchenko, S., Chowdhury, T. M. R., & Law, M. J. H. B. S. E.-P. of International Environmental. (2012). Principles of international environmental law. In S. Alam & K. Carpenter (Eds.), *Routledge handbook of international environmental law* (pp. 43–60). Abingdon: Routledge. https://doi.org/10.4324/9780203093474.ch3

Le Moli, G. (2020). The human rights committee, environmental protection and the right to life. *The International and Comparative Law Quarterly, 69*(3), 735–752.

Martin, S. (2010). Climate change, migration, and governance. *Global Governance: A Review of Multilateralism and International Organizations, 16*(3), 397–414. https://doi.org/https://doi.org/10.1163/19426720-01603008

Mayer, B. (2016). *The concept of climate migration: Advocacy and its prospects*. https://doi.org/10.4337/9781786431738

McAdam, J. (2012). *Climate change, forced migration, and international law*. Oxford University Press. https://doi.org/10.1093/acprof:oso/9780199587087.001.0001

McAdam, J. (2019). The global compacts on refugees and migration: A new era for international protection? *International Journal of Refugee Law, 30*(4), 571–574. https://doi.org/10.1093/ijrl/eez004

McAdam, J. (2020). Protecting people displaced by the impacts of climate change: The UN human rights committee and the principle of non-refoulement. *American Journal of International Law, 114*(4), 708–725. https://doi.org/doi:10.1017/ajil.2020.31

Mcadam, J., Burson, B., Kälin, W., & Weerasinghe, S. (2016). *International law and sea-level rise: Forced migration and human rights.* Lysaker: Fridtjof Nansen Institute.

Newmark, R. (1993). Non-refoulement run afoul: The questionable legality of extraterritorial repatriation programs. *Washington University Law Review, 71*(3), 833–870.

O'Riordan, T., Cameron, J., & Jordan, A. (Eds.). (2001). *Reinterpreting the precautionary principle*. London: Cameron May.

Poon, J. (2018). Drawing upon international refugee law: The precautionary approach to protecting climate change-displaced persons. In A. Kent & S. Behrman (Eds.), *'Climate refugees': Beyond the legal impasse?* (pp. 157–171). Abingdon, Oxon and New York, NY: Routledge.

Rigaud, K. K., Sherbinin, A. de, Jones, B., Bergmann, J., Clement, V., . . . Midgley, A. (2018). *Groundswell – Preparing for internal climate migration.* Washington, DC: The World Bank. https://doi.org/doi.org/10.7916/D8Z33FNS

Robinson, N. (1997). *Convention relating to the status of refugees: Its history, contents and interpretation.* New York: Institute of Jewish Affairs, World Jewish Congress.

Schlosberg, D., & Carruthers, D. (2010). Indigenous struggles, environmental justice, and community capabilities. *Global Environmental Politics, 10*(4), 12–35. https://doi.org/10.1162/GLEP_a_00029

Scott, M. (2014). Natural disasters, climate change and non-refoulement: What scope for resisting expulsion under articles 3 and 8 of the European convention on human rights. *International Journal of Refugee Law, 26*(2), 404–432.

Shacknove, A. E. (1985). Who is a refugee? *Ethics, 95*(2), 274–284.

Sharpe, M. (2018). *The regional law of refugee protection in Africa.* Oxford: Oxford University Press.

Shue, H. (1980). *Basic rights: Subsistence, affluence, and U.S. foreign policy.* Princeton, NJ: Princeton University Press.

UN General Assembly. (1972, 15 December). *United Nations conference on the human environment.* A/RES/2994. Retrieved from www.refworld.org/docid/3b00f1c840.html [accessed 17 September 2021].

UNHCR. (2020). *Procedural standards for refugee status determination under UNHCR's mandate.* Geneva: UNHCR.

United Nations. (1992, August, 12). *Report of the United Nations Conference on Environment and Development. A/CONF.151/26 (Vol. I).* Retrieved September 17, 2021, from www.un.org/en/development/desa/population/migration/generalassembly/docs/globalcompact/A_CONF.151_26_Vol.I_Declaration.pdf.

U.S. Energy Information Administration. (2020). *Country analysis executive summary: China. Independent statistics & analysis.* Washington, DC: U.S. Energy Information Administration.

Von Moltke, K. (1987). The vorsorgeprinzip in west German environmental policy. In *Twelfth report of the royal commission on environmehtal pollution.* London: H.M. Stationary Office.

Wewerinke-Singh, M. (2018). Climate migrants' right to enjoy their culture. In A. Kent & S. Behrman (Eds.), *'Climate refugees': Beyond the legal impasse?* (pp. 194–213). Abingdon: Routledge.

World Commission on Environment and Development. (1987). *Our common future.* Oxford and New York: Oxford University Press.

Young, I. M. (2006). Responsibility and global justice: A social connection model. *Social Philosophy and Policy, 23*(1), 102. https://doi.org/10.1017/S0265052506060043

3 Legal proposals and ongoing initiatives to fill the legal gap

Over the years, many proposals have been advanced to fill the legal gap[1] surrounding the figure of "climate refugees." So far, scholars have proposed the following solutions:

- a stand-alone universal treaty-based solution (Docherty & Giannini, 2009; Hodgkinson, Burton, Anderson, & Young, 2010; Prieur et al., 2008);
- regional-based solutions, including bilateral agreements (McAdam, 2011; Williams, 2008);
- piggy-backing on the international environmental regime (Biermann, 2018; Biermann & Boas, 2008, 2010; Kent & Behrman, 2018);
- the extension of the mandate of the UN refuge regime to include climate refugees[2];
- the extension of the mandate and funding of the UNHCR (Ferracioli, 2014);
- the combination of existing legal tools with a new multilateral treaty and complementary measures (Pèrez, 2018).

Despite the undeniable creativity efforts by scholars, none of these proposals have been taken up. The legal gap remains, including its "side-effects." In 2020, new displacements amounted to ca. 40.5 million, with disasters responsible for over three times compared to conflicts and weather-related events causing 98 percent of all disaster displacement (IDMC, 2021).

In the face of this global challenge, the legal impasse that surrounds the climate-induced migration as a phenomenon in itself has not, however, prevented some states from taking measures to tackle this issue. The core idea is that the absence of a legally binding definition of "climate refugees" should not preclude the possibility of developing specific policies to protect people who are forced to move in the context of climate change. Along this line of thinking, three ongoing initiatives have been taken so far: Nansen Initiative, Peninsula principles, and Kiribati's "Migration With Dignity" (MWD) relocation scheme. These initiatives largely differ in scope, ontologies, states and relevant stakeholders involved. They share, however, the same goal of filling the policy and operational protection gap.

The next paragraphs will proceed by critically analyzing legal proposals and ongoing initiatives to cope with the pressing issue of climate-induced migration.

DOI: 10.4324/9781003102632-3

Three proposals for a new universal treaty

A first strand of research has explored the possibility to fill the legal gap by laying down a draft for a stand-alone, universal Treaty dealing with climate refugees. In this respect, three proposals worth to be mentioned: a Proposal for a Convention on Climate Change Refugees (Docherty & Giannini, 2009), a draft Convention on the International Status of Environmentally-Displaced Persons (Prieur et al., 2008), and a Convention proposal for climate change displaced persons (CCDPs) (Hodgkinson et al., 2010).

The different titles adopted already suggest that conceptual obstacles remain unsettled questions. These latter determine since from the outset the different scopes of these drafts as well as authors' different positions on how to approach the issue of climate-induced migration.

The proposal advanced by Docherty and Giannini, for example, differs from the others for its clear standpoint in favor of the term "refugee." As clarified in Chapter 1 of this volume, the authors define a climate change refugee as "an individual who is forced to flee his or her home and to relocate temporarily or permanently across a national boundary as the result of sudden and gradual environmental disruption that is consistent with climate change and to which humans more likely than not contributed" (Docherty & Giannini, 2009, p. 361).

By this definition, the authors limit the scope of their convention proposal to those who cross the border because of anthropogenic climate change, thus excluding internally displaced persons as well as those who flee nature-made disruptions. Also, while their terminological choice reveals their will to frame those fleeing due to the impacts of climate change into the refugee realm, the choice of an independent convention as a legal tool uncovers the further importance they place on the need to adopt a new holistic and interdisciplinary approach. The main reason is that, according to Docherty and Giannini, this issue requires the interplay of many sources of law, including but not limited to human rights and international environmental law frameworks.

Thus, on the one hand, the new instrument envisioned by the authors adapts elements of the existing refugee definition to fit the specific circumstances of climate change. It does so, by placing more importance on the humanitarian goal of protecting victims than strict legal causation standard. On the other hand, it merges principles from environmental, human rights, and international humanitarian law. The new funding scheme the authors propose, for example, is based on international contributions pursuant to the international environmental law principle of common but differentiated responsibilities.

Further, the inclusion of social, economic, and cultural rights in the list of protections usually reserved to host state nationals only has to do with human rights and humanitarian law. Another interesting aspect of this proposal is the group-based climate change refugees status determination.

Although the authors still allow for an individual status determination in case some individuals decided to move in anticipation of environmental harm, they establish the *prima facie* recognition as a standard refugee status determination procedure for climate change refugees. The authors assume that environmental disruptions are more likely to trigger a mass influx of refugees. Group status determination should be conducted through the advisory of a body of scientific experts. This body should determine whether the environmental disruption at stake is consistent with the "climate change and more likely than not contributed to by humans" criterium (Docherty & Giannini, 2009, p. 375). Scientific experts should also provide information on states' contributions to climate change pursuant to the global fund mechanism principle of common but differentiated responsibilities for assisting climate change refugees.

Of a different kind is the universal treaty-based solution advanced by Prieur et al.

The draft convention on the international status of environmentally displaced persons was elaborated by the Interdisciplinary Center of Research on Environmental, Planning and Urban Law (CRIDEAU) and the Center of Research on persons rights (CRDP), thematic teams of the Institutional and Judicial Mutations Observatory (OMIJ), from the Faculty of Law and Economic Science, University of Limoges, with the support of the International Center of Comparative Environmental Law (CIDCE). Since from the outset, it places the issue of "climate refugees" outside the refugee realm. As recently reaffirmed by Prieur,[3] the lack of political will and legal obstacles make the option of amending the Geneva Convention to extend refugee status to environmentally displaced persons very unlikely. The scant consideration of collective rights and the protection provided only to those who cross a national boundary do not fit the case of people fleeing the effects of climate change, who are affected at collective rather than individual level and rarely cross the border (Prieur, 2018).

For this reason, the Convention opts for the broader term "Environmentally-displaced persons," defined as "individuals, families and populations confronted with a sudden or gradual environmental disaster that inexorably impacts their living conditions and results in their forced displacement, at the outset or throughout, from their habitual residence and requires their relocation and resettlement" (Prieur et al., 2008, p. 397).

Its scope is also much broader than Giannini and Docherty's proposal as it covers inter-State environmental displacements as well as internal displacements, and explicitly recognizes collective rights (e.g., the specific rights of families and populations). Further, compared to other proposals, this draft convention includes all climate scenarios: sudden/gradual environmental disasters which cause temporary/permanent displacement in the aftermath of a nature/human made catastrophe.

In this respect, the Prieur et al. proposal is truly holistic and universal as it provides tailored solutions for almost all situations "environmentally displaced persons" might find themselves.

Indeed, rights guaranteed to "environmentally displaced persons" by the Convention include not only those listed in International humanitarian law and international human rights law (e.g., right to work, health, housing, water and to food aid, etc.) but also new rights tailored to temporarily and permanently displaced persons. Examples of rights of temporarily displaced persons include: the right to safe shelter, conceived as the right to be sheltered in provisional housing established and maintained by States Parties with full respect for human dignity; the right to reintegration in their normal place of residence; the right to return to his or her normal residence; the right to prolonged shelter. Innovative rights are provided for permanently displaced persons. In addition to the right to resettlement, they also have the right to nationality, conceived as the right to conserve their home country nationality while acquiring the nationality of the receiving State.

Likewise, principles employed are those from international environmental law (e.g., principle of common but differentiated responsibilities), international law (e.g., principle of solidarity, principle of proportionality, principle of effectiveness), but also innovative principles such as that of proximity. According to the proximity principle, "environmentally-displaced persons" should be relocated in places as close as possible to their cultural area.

As in the proposal by Docherty and Giannini, the authors of this draft Convention have also envisioned a body of experts for determining the status of "environmentally-displaced persons." In this regard, 21 experts in the fields of human rights, environmental protection and peace make up the High Authority that provides assistance to the World Agency for Environmentally-Displaced Persons (WAEP).[4]

The funding mechanism is based on establishing of a designated fund called World Fund for the Environmentally-Displaced (WFED). Whereas the first funding source does not differ so much from other proposals as it derives from voluntary contributions from States and private actors, the second is unclear and somehow not convincing. According to the draft proposal, it comes from "mandatory contributions funded by a tax based principally on the causes of sudden or gradual environmental disasters susceptible of creating environmental displacements" (Prieur et al., 2008, p. 403).

The Convention proposal for climate change displaced persons (CCDPs) first differs from other proposals as it does neither apply to climate change refugees nor to all environmentally displaced persons. This Convention proposal only applies to the subset of climate change displaced persons, defined as "groups of people whose habitual homes have become – or will, on the balance of probabilities, become – temporarily or permanently uninhabitable as a consequence of a climate change event" (Hodgkinson et al., 2010, p. 90). As for a "climate change event," the proposal proceeds by specifying that this includes both sudden and gradual environmental disruption that is consistent with anthropogenic climate change. The Convention proposal for CCDPs is therefore less inclusive than that proposed for environmentally displaced persons as it is likely to exclude nature made harm. Protection and assistance are ensured to both internally displaced and those who cross international borders if they fulfill the requirements of the proposed definition, and if an international process of status designation, informed by scientific studies, affected communities, states and international institutions, recognizes them as entitled to protection.

Like Docherty and Giannini, the authors of this proposal hold that the UNFCCC is not a suitable framework for dealing with climate change displacement as its institutions are not designed to address displacement and related-issues. For this reason, they propose to establish a new institution called Climate Change Displacement Organisation (CCDO). It is made up of four bodies: an Assembly, a Council, a Climate Change Displacement Fund and a Climate Change Displacement Environment and Science Organisation (CCDESO).

Like other proposals, the group led by Hodgkinson envisions a scientific body as well as a funding mechanism based on the principle of common but differentiated responsibilities. Pursuant to this principle, developed state parties are required to make mandatory financial contributions to the Climate Change Displacement Fund (CCDF). State parties' emissions levels are monitored by the scientific body (CCDESO) that advises the Fund to determine State parties' financial contributions. The role of CCDESO is also crucial in identifying the type of climate change event driving the displacement suffered by individuals (i.e., sudden/slow onset events), thus influencing decisions made by the CCDO Council concerning requests for internal and international resettlement assistance and the level and terms of that assistance. Indeed, determinations of CCDO Council are informed by advisory provided by CCDF and CCDESO (Hodgkinson et al., 2010, p. 94).

Further, CCDO Council decision making of regional matters is expected to be informed also by regional Council committees. These latter are also involved in concluding bilateral displacement agreements between small island states and host states.

Indeed, this convention contemplates a special room for governing the peculiar situation of small island states through the principles of proximity, self-determination and the safe-guarding of intangible culture.[5] Given that small island states risk of becoming uninhabitable due to effects of climate change, authors of this convention have ensured a differentiated treatment for such a discrete population. As such, preferences of this population based on existing migration patterns, proximity, and cultural autonomy will be taken into due account by regional Council committees while concluding bilateral displacement agreements. It worth mentioning that this proposal convention contemplates tailored solutions for all climate change displaced persons depending on the severity of environmental disruption at stake. Indeed, it envisions both the pre-emptive resettlement to those most at risk in terms of the impacts of climate change, and an expanded version of non-refoulement that prevents host states from returning climate change displaced persons to areas in which their ability to survive is under threat.

Regional- and local-based proposals: regional responses, bilateral agreements, or enhanced domestic immigration laws? The case of Finland, Sweden, and Italy

A second research strand has proposed to fill the legal gap through regional instruments, including bilateral agreements, and enhanced domestic immigration laws (Mayer, 2018; Mayer, 2016; McAdam, 2011; Williams, 2008).

In this regard, some scholars argue that developing regional responses to climate change migration, perhaps under an international framework, may be more effective than having an all-encompassing treaty for various reasons. Empirically, those who move due to climate change are more likely to do so within their home or neighboring countries. Further, a regional approach would better respond to local communities' needs with tailored "solutions" accommodating particular geographical, demographic, cultural, and political circumstances (McAdam, 2011).

Angela Williams is among the first authors suggesting filling the protection gap of climate change refugees by promoting a regional system in the frame of a post-Kyoto agreement (Williams, 2008). In her view, a universal agreement should only recognize that climate change displacement exists while its management should be left in the hands of regional groupings. Such a regional system seems more suitable to respond to this pressing issue. Most regional associations are already well established, more familiar with the context-specific matters, and more equipped to take action regionally without relying on any top-down legal frameworks (Williams, 2008). Although numerous developments have significantly changed the international environmental law landscape, her critique of a stand-alone, universal Treaty as the best solution is still valid.

In a similar line of reasoning, McAdam also argues that regional and bilateral agreements can better provide tailored and targeted "solutions" for such humanitarian issues. In particular, she dwells upon the greater capacity of those instruments to respond swiftly and effectively to various climate scenarios, especially in specific geographical contexts, while considering communities' needs and willingness to leave (McAdam, 2011). On top of that, those are more likely to be adopted than multilateral treaties requiring consensus from numerous States. Thus, regional and bilateral agreements seem a more feasible and appropriate "solution" that would facilitate "economic" migration opportunities to ease problems concerning scarce resources, overcrowding, and environmental degradation before they amount to irreparable harm (McAdam, 2011).

Bilateral agreements have been often used in the Pacific Islands to secure migration pathways toward neighboring countries: Australia and New Zealand. Among the most migration-friendly strategies, it is worth noting the Pacific Access Category (PAC). Established by New Zealand in 2002, PAC allows migrants from Pacific islands of Kiribati, Fiji, Tuvalu and Tonga to move to New Zealand every year in compliance with existing quotas. So far, this circular migration scheme has allowed some beneficiaries to settle in New Zealand permanently. Indeed, as soon as they find a job, a permanent residence status can be obtained (Klepp & Herbeck, 2016).

However, bilateral agreements risk being used by some States to take advantage of pre-existing unequal power relations, thus resulting in unfair outcomes. A useful example is the 2017 Italy-Libya Memorandum of Understanding signed between the then Italian Prime Minister Gentiloni and Fayez al-Serraj, Head of the UN-backed Libyan Government of National Accord,[6] which resulted in increased smuggling and reinforcement of border security. Rather than filling the protection gap for those fleeing from the Sub-Saharan region, this policy approach aims at reducing crossings from Libya to Italy at any cost. This Memorandum does not explicitly target climate-induced migrants. Nevertheless, one can reasonably argue that those people are indirectly included as they are more likely to end up in mixed flows of refugees, vulnerable migrants, and people displaced by climate change and disasters. Indeed, the World Bank reported that Sub-Saharan Africa is among the most affected world regions by climate change, with 85.7 million climate migrants expected to move by 2050 (Clement et al., 2021). Against this background, such Memorandum falls within the frame of pre-existing unequal power relations built upon the numerous agreements focused on curbing migratory flows and facilitating readmission since the 2000s and throughout the Gaddafi regime.

Similarly, regional agreements can be more effective than multilateral instruments in practical terms. However, they fail to address the ethical issues linked to the greater responsibilities developed states have compared to developing countries in creating the root causes of climate-induced migration. Given that most climate migrations/displacements stem from the Global South, a regional solution could place the burden of resettlement/humanitarian assistance on the shoulders of the most vulnerable/less equipped countries. Such a double standard particularly emerges from the European legal framework. When a disaster occurs, the free movements of people provisions enshrined by the EU treaties can be used only for cross-border movement between the EU-Member States. Since 2020, the only suitable instrument for non-EU citizens has been the Temporary Protection Directive 2001/55/EC.[7] This instrument aimed at addressing "refugee-like" situations of people fleeing violence, and systematic or generalized violations of their human rights. However, although potentially applicable to environmentally displaced people from outside the EU, it has never been used to protect them. On 23 September 2020, the New Pact on Migration and Asylum[8] presented by the EU Commission repealed the Directive 2001/55/EC by replacing the "temporary protection" with the "immediate protection" regime for cases of mass influx. While we wait for the possible applications of this new regime to large-scale climate displacement, one can observe the lack of inclusion of an explicit category of environmental migrants. Not only remains the New European Pact on Migration silent on this pressing issue, but it excludes the possibility to create new subjects deserving to be qualified as international refugees outside the scope of the 1951 Refugee Convention.

Proposals to enhance domestic immigration laws do not seem promising too. If we look at the EU-Member States, their policies are mostly oriented toward border control through measures to prevent landings, and curb migration flows. Not surprisingly, domestic immigration laws are more likely to be repealed with narrower rather than broader scope. In this regard, a useful example can be found in Swedish and Finnish migration laws.

Until 2015/2016, Finland and Sweden were the only countries in the EU providing specific humanitarian protection to environmental migrants (Hush, 2017; McAdam, 2020). In particular, Section 88a, Chapter 6 of the Finnish Aliens Act 301/2004 granted humanitarian protection if an environmental catastrophe prevented a person from returning home. In that case, a Finnish resident permit could be issued if neither asylum nor subsidiary protection was applicable. Similarly, Section 2a, Chapter 4 of the Swedish Aliens Act (2005: 716) ensured protection to a person who could not return home because of an environmental disaster. Although this provision was relevant in ca 170 claims between 2006 and 2016, it has never resulted in a single grant of status.[9]

However, because of the high number of arrivals in 2015, both countries suspended their domestic provisions. While Finland repealed those provisions in 2016, Sweden opted for a temporary repeal from 20 July 2016 to 19 July 2019, subsequently extended (in June 2019) until July 2021. The 2021 reform of the Swedish Aliens Act finally removed those provisions while amending numerous rules of the former Aliens Act (2005: 716),[10] so that, to date, environmental migrants do not have the right to a residence permit in Sweden.

At the time of writing, Italy would seem the only exception as for national migration laws granting humanitarian protection to environmental migrants. The Decree-Law of October 21, 2020, No. 130, converted with amendments into Law December 18, 2020, No. 173 enshrines that residence permits for people fleeing disasters can be converted to longer-term work visas.

Indeed, while Finland and Sweden suspended their domestic forms of humanitarian protection to environmental migrants, in 2015, the Tribunal of Bologna (Italy) ensured humanitarian protection to Rachid: an immigrant from Pakistan who came to Italy in the aftermath of a flood. His lawyer, Alba Ferretti, obtained this favorable judgment (later confirmed on appeal) on the basis of "serious reasons" of a humanitarian nature as laid down in Article 5 (6) of Legislative Decree No. 286/1998 (Italian Immigration Act).[11] For the first time in Italy, humanitarian protection was granted for environmental reasons. In practical terms, the plaintiff (i.e., Rachid) obtained a residence permit then converted to a longer-term work visa. In that case, Article 5 (6) of the Italian Immigration Act was conceived as a "safeguard clause" so inclusive that it covered even environmental migrants excluded from any protection until then.

Not surprisingly, a few years later, the 5 Stars-League government repealed the reference to the "serious reasons" of a humanitarian nature through Decree No. 113/2018 (also known as "Security Decree"). The basic assumption was that the lack of a legal classification of such "serious reasons" of a humanitarian nature was increasing the number of beneficiaries significantly. For this reason, the "Security Decree" dismantled humanitarian protection in the first place and then replaced it with seven temporary residence permits for humanitarian purposes, including disasters.

"The residence permit for disasters," however, could be eventually renewed for another six months only if exceptional disaster-related conditions persisted in the applicant's country of origin, while the right to receive a longer-term work visa was repealed. This provision made the humanitarian protection substantially inapplicable, thus leaving thousands of migrants into irregular status,[12] with the recognition of humanitarian protection shifted from 25 percent of cases of protection in 2017 to 1 percent in 2019.[13]

Thanks to the current Law No. 173/2020 the "residence permit for disasters" (*permesso di soggiorno per calamità*) has been significantly amended. The most relevant novelty concerns the possibility of converting residence permits for people fleeing disasters into longer-term work visas. Crucially, Law No. 173/2020 has deleted the former reference to the renewal of this residence permit for six months only if exceptional disaster-related conditions persist in the applicant's country of origin. With the new formulation, indeed, the residence permit for disasters can be renovated without any fixed term as long as the conditions of insecurity in the country of origin persist and eventually converted to a longer-term work visa.

Further relevant novelties introduced by Law No. 173/2020 have to do with the wording of Article 20-*bis* of the Italian Immigration Act, which governs the residence permit for disasters. In particular, amended Article 20-*bis* has replaced the wording "imminent and exceptional" disaster (*contingente ed eccezionale*) with "serious" disaster (*grave*), thus allowing a potentially broader interpretation of the type of disaster to be considered. According to some legal scholars (Di Pietro, 2021; Scissa, 2021), a "serious" disaster may well include both slow and rapid onset events as the emphasis is given to the severity rather than the imminence of the climate event. Further, the term "disaster" is still left undefined, so that it remains unclear whether it can include both natural and human-made disasters. In this regard, most legal experts argue that the term "disaster" is more likely to be interpreted as natural and human-made in compliance with Article 1 of Legislative Decree No. 1/2018, which defines disaster in such twofold understanding (Benvenuti, 2019; Di Pietro, 2021; Scissa, 2021).

With ordinance No. 5022/2021,[14] the Court of Cassation further clarified that the recognition of humanitarian protection is justified in any context which poses a real risk to fundamental rights to life, freedom, self-determination of individuals, including climate change and unsustainable use of natural resources. The Court also emphasized the concept of the inviolable core of human dignity as a prior condition for granting humanitarian protection. Ultimately, the risk posed to the inviolable core of human dignity upon repatriation and the subsequent individual vulnerability legitimate the recognition of humanitarian protection. In practical terms, humanitarian protection can be granted to environmental migrants – who do not qualify for international protection – if a serious disaster in their country of origin prevents them from returning home safely.

However, what Law No. 173/2020 has not amended is the authority responsible for assessing the severity of the disaster. Like in the former formulation of Article 20-*bis*, the residence permit of disasters is still issued at the discretion of the police headquarters (*Questura*). Such discretion has been a focus of a challenging debate following a 2021 Field Monitoring[15] in 15 Italian cities reporting the failure to comply with new provisions introduced by Law No. 173/2020. The study shows that police headquarters are still applying the repealed Security Decree while disregarding, delaying, and ultimately boycotting requests for special protection. Circular No. 23186 of 19 March 2021, issued by the Ministry of the Interior, has further complicated the application procedure for residence permits. In this regard, parliamentary question No. 4–05444 presented by Senator Gregorio De Falco[16] reported the introduction of numerous unnecessary bureaucratic hurdles, such as presenting applications for the issue or renewal of a residence permit for special protection in person or by mail only. This provision excludes the possibility of presenting such applications via e-mail, which would be easier, faster, and safer at the time of Covid-19.

As a result, unlawful practices by police headquarters, bureaucratic hurdles, and contradictory provisions introduced by Circular No. 23186 are making new provisions introduced by Law No. 173/2020 *de facto* inapplicable.[17]

To conclude, although the residence permit for disasters is a remarkable form of protection to "environmental migrants," the discretion of police headquarters is vanishing affecting its effective implementation. Further, it remains unclear whether it is a "real" step forward in the Italian legislation on this matter or rather two steps back compared to the broader humanitarian protection provided by the repealed Article 5 (6) of Italian Immigration Act.

Beyond silos: connecting different international law regimes

Following Williams' proposal, other scholars pushed the argument of installing the issue of "climate refugees" in the frame of the international environmental regime even forward in light of recent environmental law developments. The basic assumption is that the issue of "climate refugees" can be better addressed by principles and practices of international environmental law in the frame of global climate governance (Biermann, 2018; Kent & Behrman, 2018). Not only has the Paris Agreement offered the necessary overarching legal framework, thus avoiding risky negotiations for creating a stand-alone treaty, but it has also established relevant funding mechanisms. Indeed, in recognizing the major developed states' responsibilities concerning the root causes of climate-induced migration (e.g. CO_2 emissions), the Paris Agreement has already introduced useful mechanisms for seeking redress and compensation. A key example is the creation of the 2015 Task Force on Displacement under the aegis of the Warsaw International Mechanism (WIM), established in 2013 (also) to channel financial support for addressing loss and damage associated with the adverse effects of climate change. On top of that, the climate change regime offers different financial mechanisms such as those provided by the Kyoto Protocol and dedicated environmental funds like the Global Environmental Facility (GEF) and the Green Climate Fund (GCF) (Kent & Behrman, 2018).

In light of these considerations, two proposals are worth mentioning. The first has been advanced by Biermann and Boas in 2008, then slightly amended by Biermann in 2018 (Biermann, 2018; Biermann & Boas, 2008), and suggests adding a UNFCCC Protocol to provide for a *sui generis* regime for the recognition, protection, and resettlement of "climate refugees." According to the most recent version, the institutional development of this regime should be built upon five principles: *planned re-location and resettlement; resettlement instead of temporary asylum; collective rights for local populations; international assistance for domestic measures; and international burden-sharing* (Biermann, 2018). The core idea is that, in the long term, emergency response and disaster relief should be replaced with planned and voluntary resettlement. Indeed, "climate refugees" differ ontologically from "political refugees" as they are not expected to return to their countries once the conflict is over. Environmental degradation at later stages poses serious threats to the life of entire communities, thus causing irreversible harms that do not allow people to return.[18] Accordingly, such a *sui generis* regime should:

1 conceive "climate refugees" as permanent immigrants to host countries;
2 design provisions tailored for collectives of people;
3 provide international assistance and funding for domestic measures given that most climate change impacts affect specific regions within countries; and
4 establish fair funding mechanism imposing most developed (and responsible) countries to bear higher costs for protecting climate refugees (in compliance with the common but differentiated responsibilities principle).

Such UNFCCC Protocol should also include an executive committee supported by a specialized scientific body and a *sui generis* financial support mechanism: a Climate Refugee Protection and Resettlement Fund. As clarified by Biermann, this latter should be conceived as an independent funding mechanism having new and additional funds to avoid competition over other existing funds (Biermann, 2018).

Finally, Biermann calls for early action for establishing appropriate governance mechanisms through dedicated institutional settings. As explicitly mentioned in the first iteration of this proposal (Biermann & Boas, 2010), an appropriate governance would imply the creation of a network of implementing agencies under the umbrella of the UNFCCC Protocol. At a minimum, such network should include the UN Development Programme (UNDP) and the World Bank, while other policy making actors to be included might be the United Nations High Commissioner for Refugees (UNHCR), the United Nations Environment Programme (UNEP), the World Health Organization (WHO), or the UN Food and Agriculture Organization (FAO). In their view, this governance architecture aims to incorporate measures to combat climate change impacts into agencies' programs. Since they propose this inter-institutional perspective, many step forwards have been taken by major international agencies committed with migrants/refugees/displaced people toward the construction of a cross-governance model. Of particular importance are the two systems of governance created in 2015 under the IOM (i.e., its Division on Migration, Environment and Climate Change, MECC) and the UNFCCC (i.e., its Task Force on displacement, TFD) institutional settings.[19] Both systems are somehow realizing what Biermann and Boas anticipated in their proposal as they have created institutional settings for sharing expertise and practices from the most relevant stakeholders active in the field of climate-induced migration and displacement.

The second proposal is Kent and Behrman's piggy-backing approach to the international environmental regime (Kent & Behrman, 2018). The authors justify such a piggy-backing on the international environmental regime on the following grounds. First, they recognize the greater feasibility of encapsulating climate refugees' recognition within the existing climate change regime rather than opening new negotiations on the refugee convention, more likely to provide for narrower than broader protection mechanisms. Second, they point out the greater suitability of international environmental law to address the climate migration phenomenon over human rights law. In particular, they dwell upon the overly individualized approach of human rights and refugee law that prevents from adequately addressing the peculiar violence caused by climate change. The type of violence described by the refugee Convention focuses on persecution against single individuals. However, such individual-based persecution ontologically differs from the violence of climate change that hits large and potentially indiscriminate groups of people. Above all, the authors argue that such an individualized approach makes it challenging to identify duty bearers and right holders when it comes to the nebulous effects of climate change.

Unlike Biermann and Boas, the authors do not believe that creating an additional Protocol is necessary. Rather, they extensively pursue the idea of overcoming the "legal impasse" through joint efforts of existing laws. In their view, human rights, refugee and environmental laws should be seen as complementary regimes for governing the complexity of climate-induced migration under what they call a "cross-governance model": a "formal, accommodating institutional setting, in which actors from different institutions, areas of expertise and perspective can operate (and cooperate) in a coordinated manner" (Kent & Behrman, 2018, p. 152).

In this context, environmental law regime provides the most appropriate framework for this cooperation as it is more suitable to capture the collective-based "violence" of climate change, and it is more financially capable than human rights and refugee regimes.

Thus, an important contribution to the lively debate on how to fill the legal gap surrounding climate-induced migration relies on their proposal to base future agreements on collective rights and states' obligations. Notably, Kent and Behrman have elaborated a "hot-spots"-based collective rights model where states, rather than individuals, are considered as both 'rights holders' and "duty bearers." Not only is it more practical identifying states most affected by climate-induced migration ("hot-spots"), but it is also more feasible convincing states to financially support and assist receiving states. Unlike human rights and refugee regimes, existing financial mechanisms under the UNFCCC are already inspired by the common but differentiated responsibility principle and respective capabilities. Indeed, as reaffirmed by Article 9 of the 2015 Paris Agreement, developed country Parties are to provide financial resources to assist developing country Parties in implementing both mitigation and adaptation objectives of the UNFCCC. In light of this, the authors argue that the UNFCCC is the appropriate home for regulating this phenomenon because of its multilateral nature, legal principles such as common but differentiated responsibilities, and increased financial resources (Kent & Behrman, 2018).

Finally, the authors further justify such a centrality of states on the basis of empirical evidence showing that climate-induced migrants rarely cross borders as they more often move within their countries. Therefore, it would make more sense to have States as both "rights holders" (as they are expected to host, or already are hosting climate migrants/refugees) and "duty bearers" (as polluting states are most responsible for having caused climate change impacts such as migration).

I agree with the authors that a new legal instrument might not be the answer and that more efforts should be devoted to bolstering cross-fertilization across different legal silos. However, I can't entirely agree with the subjects they identified as duty bearers and rights holders. In Chapter 4, I lay out an argument as to why such a state-centric approach should be replaced with a non-state-centric one while trying to provide a way forward.

Just a matter of extension?

Authors engaged with the topic of "climate refugees" also disagreed on whether we need a new instrument whatsoever or it is possible to protect those people on the move through an extension of the existing refugee regime.

As reported by Biermann and Boas (Biermann & Boas, 2008, 2010), the Republic of the Maldives proposed to broaden the mandate of the 1951 Convention and of the UNHCR to cover also "climate refugees."[20] Such an extension has been widely questioned by Ferracioli, who has investigated at length the feasibility of broadening the scope of the refugee Convention as well as the mandate and funding of the UNHCR (Ferracioli, 2014). The author highlights that before opening negotiations for adopting a new and broader Convention, motivational and institutional constraints should be alleviated in the first place. In her view, these consist of the concern with diplomatic relationships between the host country and the country of origin, the unauthorized entrance of immigrants into the state territory, and the adverse public opinion. Having accomplished that, the author argues that the collective agents most suitable to both improve public's attitude toward asylum seekers and the global governance of migration and refugee protection are most powerful states and the UNHCR. To achieve these goals, she suggests that those actors should shift the emphasis from integration to resettlement in order to make the public attitude's more sympathetic to the refugees. The main argument is that resettlement is already an accepted solution on part of the international community, so that it does not require the negotiation of a new Convention. However, this shift would certainly require an expansion of the mandate and funding of the UNHCR to provide clear legal guidance and sufficient material assistance to each state for processing those in its territory and resettling those processed elsewhere (Ferracioli, 2014).

Importantly, Ferracioli has shown that positive law cannot fill such a legal gap as long as institutional and motivational constraints are not alleviated before opening the negotiations for a new Convention. Although I agree that the law-creation option to fill this legal gap is more likely to fail the "feasibility test," I would rather turn the focus of this debate on overcoming the legal gap through analogy.

As anticipated in Chapter 1, this option first implies demonstrating that two cases are alike as they meet the standard of comparison: the underlying purpose for a specific norm (i.e., the *ratio legis*). Second, once the similarity has been established, it justifies the inference of the same legal consequences on the ground that, if two cases are relevantly similar, they ought to be treated alike (Juthe, 2016).

As for the case of "climate refugees," I argue that one can easily find similarities between political and climate refugees by looking at the same vulnerable situations they face. Given that the underlying purpose for the international refugee law is providing international assistance to those deprived of basic rights with no recourse to home government (Shacknove, 2010), one can come to the following conclusion. Anytime the fundamental human rights of migrants in vulnerable situations are at risk upon return to their home countries, those people should be treated alike regardless of the reasons why they escape from their countries of origin (e.g., persecution, war, natural disasters). It follows that, once such a legal analogy is established along the line of vulnerability, it is likely to trigger the non-refoulement principle as protection has to be equally granted to "climate refugees." The concept of vulnerability includes different nuances that allows us to identify climate "hot spots," disadvantaged communities, and "low-mobility" individuals. Not only climate "hot spots" can be used as readily apparent circumstances to provide for a *prima facie* approach, but they can also serve as an empirical basis to trace migration flows. Although not legally binding, the Global Compact on Refugees has acknowledged that climate change, environmental degradation, and natural disasters increasingly interact with the drivers of refugee movements.[21] As a result, large movements can also include forcibly displaced people fleeing sudden-onset natural disasters and environmental degradation. To sum up, I argue that a legal analogy drawn along the line of vulnerability may well pave the way for overcoming such a "legal impasse." I will develop this argument in more detail in Chapter 4 of this volume.

Combining existing legal framework with new multilateral treaty and complementary measures

A further strand of research consists of combining existing legal instruments with a new multilateral treaty depending upon the type and severity of environmental disruptions (Pèrez, 2018).

Advanced by Felipe Pérez, this proposal challenges the idea that an all-encompassing treaty is the only game in town to "solve" the legal impasse. By contrast, she identifies four categories of climate migration and provides tailored "legal solutions" for each. In her view, climate migration is a heterogeneous phenomenon that cannot be addressed through a single branch of international law or legal instrument.

Drawing on the previous categorization provided by Kälin (Kälin, 2010), she lists four categories of climate migration:

> climate emergency migration (a scenario occurring before or in the aftermath of rapid-onset events);
> climate-induced migration (slow-onset events at their early stages);
> climate-forced migration (environmental degradation at the later stages);

migration in/from small island states (a sub-category of climate-forced migration modelled on the peculiar geographical context of Small Island Developing States, SIDS).

Those categories show that migrants' needs (of protection) largely vary from the type of environmental disruptions, thus requiring different solutions. Among those, she lists the present international law, complementary protection systems, a new multilateral treaty, and a stronger commitment by governments to implement the proposed legal protection regime. Thus, potentially applicable legal frameworks are international human rights law, international labor migration law, international refugee law, the 1998 Guiding Principles on Internal Displacement (although not legally binding), the 2009 African Union Convention for the Protection and Assistance of Internally Displaced Persons in Africa (although not all African States have ratified it), and the 1954 Convention relating to the Status of Stateless Persons. On top of that, she also mentions the necessity to adopt a binding universal treaty as a long-term option. Such a treaty should recognize a legal status for climate migrants in certain circumstances while guaranteeing all climate migrants' fundamental rights. Different vulnerabilities of climate migrants should be addressed by taking into account gender, age, health, level of education, ethnicity, and income. Like other proposals, she also points out the necessity to integrate the principle of common but differentiated responsibility into the treaty to create obligations for most polluting states to assist the most affected states financially. Although such a treaty might be the best option to deal with this phenomenon, she also concludes that this solution risks being not feasible due to the unwillingness of States to adopt and implement those new international obligations. In practical terms, such a heterogeneous phenomenon is to be tackled by using all existing instruments in place, aware that a new multilateral treaty should be part of the different solutions in the longer term.

The following paragraphs will examine what the author has termed complementary measures: Nansen Initiative, Peninsula Principles, and Kiribati's "Migration With Dignity" (MWD) relocation scheme.

Nansen initiative

The Nansen Initiative is a state-led consultative process launched in 2012 by Norway and Switzerland to reach a consensus over a protection agenda to tackle greater vulnerabilities faced by migrants in the context of climate change and disasters. Together with other States involved in this initiative, both countries recognized the lack of responses to cross-border disaster-displacement and the existence of a legal and operational gap preventing from meeting migrants' needs and vulnerabilities. In addressing this gap, the Nansen Initiative engaged a process of regional consultations with numerous stakeholders ranging from international organizations, government officials, civil society, and academics, to gather domestic and regional effective practices and approaches to face this pressing issue (The Platform on Disaster Displacement, 2018b).

Built upon the outcome of the 2011 Nansen Conference on Climate Change and Displacement[22] held in Oslo and the Cancun Agreement, these consultations resulted in adopting the *Agenda for the protection of cross-border displaced persons in the context of disasters and climate change* in 2015. More than 100 governmental delegations at the Global Consultation in October 2015 in Geneva endorsed the Agenda, thus marking the start of a new phase of this initiative (Platform on Disaster Displacement, 2018a). This new phase was inaugurated through the creation of the Platform on Disaster Displacement (PDD), a follow up on the work started by the Nansen Initiative. The PDD was launched at the World Humanitarian Summit in May 2016 with the aim of bolstering protection of cross-border disaster-displaced persons through enhanced involvement and partnership between policymakers, practitioners, and academics (The Platform on Disaster Displacement, 2018b).

The PDD has adopted a multi-stakeholder approach for facilitating the exchange and integration of effective practices into domestic and regional policies and normative frameworks. In this view, the Platform comprises a Steering Group made up of States, a Coordination Unit and an Advisory Committee of experienced individuals and representatives of international and regional organizations. Members of the Steering Group include Australia, Bangladesh, Brazil, Canada, Costa Rica, European Union, Fiji, France, Germany, Kenya, Madagascar, Maldives, Mexico, Morocco, Norway, Philippines, Senegal, Switzerland, and standing invitees such as the United Nations High Commissioner for Refugees (UNHCR) and the International Organization for Migration (IOM). Those latter are also conceived as key operational partners expected to implement the PDD 2016–2019 Strategic Framework and Workplan (The Platform on Disaster Displacement, 2018b). Based on the recommendations of the Protection Agenda, the Strategic priorities are the following:

"I   Address knowledge and data gaps;
II   Enhance the use of identified effective practices;
III Promote policy coherence and mainstreaming of human mobility challenges in, and across, relevant policy and action areas;
IV   Promote policy and normative development in gap areas" (The Platform on Disaster Displacement, 2018b, p. 423).

These priorities are set under the leadership of the Member States, who have significantly shaped the terminology while reframing the debate on climate-induced migration and displacement within the need to reduce disaster risks and provide humanitarian protection measures.

In PDD's perspective, the disaster plays a key role in shaping International Climate Politics on climate migration and displacement. Following the UN International Strategy for Disaster Reduction (UNISDR), a disaster is defined as a "serious disruption of the functioning of a community or a society involving widespread human, material, economic or environmental losses and impacts, which exceeds the ability of the affected community or society to cope using its own resources" (Platform on Disaster Displacement, 2018a, p. 129).

The point is not distinguishing between anthropogenic or natural disasters, as the main priority is to protect people in need of assistance, regardless of whether they flee a tsunami, an earthquake, or a flood. Not surprisingly, expressions suggesting the causal attribution of climate change-related events (e.g., "as a result of," or "due to" the effects of climate change) are rarely employed and more often replaced by more cautious expressions like "in the context of disasters and climate change."

Although this soft-law solution has the merit of taking seriously the question of cross-border disaster-displaced persons, it lacks a comprehensive understanding of the climate-induced migration in all its complexity. Indeed, it extensively focuses on disasters while neglecting other forms of environmental degradation such as slow-onset events that can equally lead to migration and displacement in the longer term. In other words, climate-induced migration is more nuanced than that, and this representation risks disregarding its bigger picture. On top of that, the terminology employed revolves around the category of displacement that does not trigger any legal obligations and mostly relies on member states' discretion in implementing PDD's recommendations.

Peninsula principles

A further complementary measure can be found in the domain of Internal Climate Displacement. In the face of evidence showing that climate-displaced persons mostly move internally and do not cross borders, grassroots organizations, international NGOs, and experts have emphasized the need to ensure their human rights through a dedicated set of principles and norms (Hassine, 2019).

With this aim, the non-governmental organization *Displacement Solutions* was established in 2006.

Under the leadership of his director, Scott Leckie, *Displacement Solutions* aims to address climate displacement and meet climate-affected persons' needs on the ground, thus working in cooperation with local NGOs, grassroots organizations, and international experts. A dedicated project, the Climate Displacement Law Initiative, was launched in 2009 to focus on actions and outputs to provide practical guidelines on climate displacement within States anchored in existing international human rights instruments. One of the most important outputs of this project was the adoption of a new international soft law instrument in August 2013: the Peninsula Principles on Climate Displacement Within States (hereinafter the Peninsula Principles).

Consistent with the 1998 UN Guiding Principles on Internal Displacement, this soft law instrument specifically addresses challenges of countries hardest hit by climate change, such as Bangladesh, Fiji, Kiribati, Tuvalu, Vanuatu, Maldives, and Papua New Guinea. In particular, it provides a "comprehensive normative framework, based on principles of international law, human rights obligations and good practice, within which the rights of climate displaced persons can be addressed."[23] Unlike other instruments discussed so far, the Peninsula Principles evolved from a five-year process of broad-based consultations with people living in most affected countries and grassroots organizations whose quest for guidance and solutions has been then integrated into this soft law instrument. The basic assumption of the Peninsula Principles is that communities are expected to play a key role throughout the climate displacement process as they will need to organize themselves, claim their rights, and outline corresponding States obligations.

The resulting normative, institutional, and implementation framework has a clear global advocacy purpose. The Peninsula Principles provide a globally relevant standard on climate displacement aimed at governments committed to legal and policy initiatives in this field.

Described as the first formal policy of its kind in the world, the Peninsula Principles certainly cover the majority of climate-displaced peoples as most of them move internally. Although not legally binding, it is an instrument that might partially fill the legal gap, especially if combined with other measures and existing legal frameworks. Hopefully, it might eventually pave the wave for a more ambitious project in the future, perhaps toward a global, legally binding Convention such as the 2009 Kampala Convention.

Migration with dignity

As anticipated in Chapter 1, Small Island Developing States (SIDS) are a particular case both for the severity of environmental degradation they are facing and innovative solutions they are using to cope with it. Given the peculiarity of their geographical, economic, and political context, they are considered the case-study *par excellence* of climate migration and displacement.

Defined as the *synecdoche* of climate change (Connell, 2003; Hirsch, 2015; Hulme, 2010), these islands (e.g., Papua New Guinea, Kiribati, Maldives, Tuvalu, Maldives, Fiji) are extremely vulnerable to sea-level rise, a phenomenon *already* experienced in the form of a slow-onset event in its later stages of environmental degradation. Such slow violence (Nixon, 2011) has already forced people living in the Carteret Islands (Papua New Guinea) to leave their habitual residences. Significant relocations to Bougainville have *already* occurred over the last years as the sea level rise has made Carteret islands uninhabitable due to overcrowding, inundation, crops, and water supplies destroyed by saltwater intrusion (Campbell, 2012; Kupferberg, 2021). Not surprisingly, the media portrayed those people forced to flee as the world's first climate refugees.[24]

The voluntary relocation of indigenous communities has been organized and facilitated by *Tulele Peisa* (Sailing the waves on our own), a local NGO based in Bougainville (United Nations Development Programme, 2016). The relocation program started in 2009. To date, it is expected to be completed far longer than what has been initially estimated. Early forecasts projected the end of the resettlement program by 2012. However, the staff at *Tulele Peisa* encountered many obstacles, especially in terms of available resources to support the program. Further obstacles derived from acquiring legal rights to land in the face of local landowners' opposition made it difficult to find adequate sites to resettle people from Cartaret Islands. On top of that, relocatees were often left without adequate support once they arrived at Bougainville. For this reason, some of them eventually decided to return to the Cartaret Islands (Kupferberg, 2021). As a result, based on updated estimates, the current intention is to move 1,350 people or 50 percent of the total population by 2020 (United Nations Development Programme, 2016).

Given the numerous obstacles and the subsequent delay throughout the process, such a resettlement scheme can be assessed as unsuccessful.

An alternative relocation scheme much debated by the literature is Kiribati's "Migration With Dignity" (MWD) (Borras, 2016; Kupferberg, 2021; McAdam, 2012). In Kiribati, the core idea underpinning this planned relocation scheme was to secure the minimally good life while retaining a sense of dignity in the face of climate change.

The former President of Kiribati, Anote Tong, employed this idea to design the migration policy known as "migration with dignity" aimed at facilitating migration through dedicated training and education for successfully adapting to the labor market and the society of destination countries. Based on the 2013 Dhaka Principles,[25] its main goal was to promote business while protecting the human and labor rights of migrant workers (Borras, 2016).

Such a migration policy consistently implied the unwillingness to be labeled as refugees, as the term could eventually undermine the national pride and constitute a sort of social stigma (Borras, 2016; Kupferberg, 2021). By contrast, Anote Tong introduced this planned relocation scheme to gradually relocate I-Kiribati (i.e., Kiribati's indigenous residents) to Fiji. In 2014, he purchased 20 sq km in Vanua Levu, one of the Fiji Islands, to provide a piece of land to I-Kiribati should the island be submerged.[26]

This scheme cannot be considered successful either. Succeeding Anote Tong in 2016, Taneti Maamau, the current President of Kiribati, replaced the Migration With Dignity relocation scheme with a faith-based approach to promote economic growth in compliance with adaptation and mitigation policies. Thus, instead of planning for the worst, Taneti Maamau is devoting most of his efforts to strengthen resilience to keep people in place. In practical terms, he believes that bringing more financial resources from international aid and fostering traditional activities (e.g., fishing, coconut trade, and tourism) is crucial to accumulate the necessary resources to tackle the high costs of climate change (Kupferberg, 2021; Walker, 2017). In other words, he does not deny that climate change is *already* affecting the whole population, but he replaced the apocalyptic "We will go down" narrative of Anote Tong with a post-apocalyptic, faith-based narrative supporting the belief that only God might eventually unmake the island (Cassegård & Thörn, 2018). As he declared, "We try to isolate ourselves from the belief that Kiribati will be drowned. . . . The ultimate decision is God's" (Walker, 2017).

However, this race against the clock driven by hope in God's benevolence and enhanced resilience has been criticized by some scholars as it might not be sufficient to tackle the effects of climate change. The case of Kiribati, along with other examples of relocation schemes in the Pacific Islands, has demonstrated that there is no one encompassing "solution" to address the impacts of climate change. Rather, the use of different approaches and measures might be more promising if consolidated in due time (Kupferberg, 2021).

Notes

1 Among the few exceptions, see Jessica Cooper's position arguing that there is no legal gap as environmental refugees already meet the requirements of the 1951 refugee (Cooper, 1997).

2 Republic of the Maldives Ministry of Environment, Energy and Water, Report on the First Meeting on Protocol on Environmental Refugees: Recognition of Environmental Refugees in the 1951 Convention and 1967 Protocol Relating to the Status of Refugees, Male, Maldives, 14–15 August 2006, cited in (Biermann & Boas, 2008).

3 Cf. a third version of the draft convention written by the Limoges group in 2013 is available at https://cidce.org/wp-content/uploads/2016/08/Draft-Convention-on-the-International-Status-on-environmentally-displaced-persons-third-version.pdf [accessed 14 May 2021].

4 See article 11 for more details on its role, main tasks, and organization.

5 Hodgkinson et al. describe the proximity, self-determination and the safe-guarding of intangible culture principles as follows. Proximity principle is retrieved by the proposal by Prieur et al. and implies that climate change displaced persons are to be relocated in places as close as possible to their cultural area. Self-determination principle adds that climate change displaced persons should choose when to leave their home countries and where to be relocated.

Finally, the safe-guarding of intangible culture principle enshrines the legal protection of climate change displaced persons' cultural autonomy. This principle would imply to enable the preservation, transmission, and revitalization of such cultural heritage, if the territory of small island states ceased to exist. In this regard, Hodgkinson et al. suggest anchoring such a mechanism of preservation under the 2003 UNESCO Convention for the Safeguarding of the Intangible Cultural Heritage.

6 The 2017 Memorandum of understanding on cooperation in the fields of development, the fight against illegal immigration, human trafficking, and fuel smuggling and on reinforcing the security of borders between the State of Libya and the Italian Republic is available in English at https://eumigrationlawblog.eu/wp-content/uploads/2017/10/MEMORANDUM_translation_finalversion.doc.pdf

7 European Union: Council of the European Union, Council Directive 2001/55/EC of 20 July 2001 on Minimum Standards for Giving Temporary Protection in the Event of a Mass Influx of Displaced Persons and on Measures Promoting a Balance of Efforts Between Member States in Receiving such Persons and Bearing the Consequences Thereof, 7 August 2001, OJ L.212/12–212/23; 7.8.2001, 2001/55/EC, available at: www.refworld.org/docid/3ddcee2e4.html [accessed 1 October 2021].

8 European Commission, Communication from the Commission on a New Pact on Migration and Asylum, COM (2020) 609 final, Brussels, 23.9.2020.

9 I want to express my gratitude to Prof. Matthew Scott for sharing this information while writing this chapter.

10 See Regeringens Proposition 2020/21:191: Ändrade Regler i Utlänningslagen, p. 51.

11 See Order of the Court of Bologna No. 7334/2014 R.G. of 17/11/2014, and Order of the Court of Appeal of Bologna n° 504/2016 of 29/03/2016.

12 According to the 2020 ISMU Migration Report, irregular migrants amounted to 562,000 at the end of 2018. The report is available at www.migrantes.it/wp-content/uploads/sites/50/2020/10/RICM_2020_DEF.pdf [accessed 7 October 2021].

13 See Facchini, D. (2021). Immigrazione: così il ministero dell'Interno vanifica il superamento dei "decreti Salvini." Retrieved October 7, 2021, from https://altreconomia.it/immigrazione-cosi-il-ministero-dellinterno-vanifica-il-superamento-dei-decreti-salvini/

14 See Court of Cassation ordinance No. 5022/2021, available at www.italgiure.giustizia.it/xway/application/nif/clean/hc.dll?verbo=attach&db=snciv&id=./Oscurate20210305/snciv@s20@a2021@n05022@tO.clean.pdf [accessed 7 October 2021].

15 The Monitoring Field titled *Paradosso all'italiana. Quando il governo italiano boicotta se stesso* [Italian Paradox. When the Italian government boycotts itself] is

available at www.percambiarelordinedellecose.eu/wp-content/uploads/2021/05/para-
dosso_maggio2021.pdf [accessed 7 October 2021].

16 Parliamentary question No. 4–05444, published on 12 May 2021, in session no. 325,
    is available at www.senato.it/japp/bgt/showdoc/showText?tipodoc=Sindisp&leg=18
    &id=1297549 [accessed 7 October 2021].

17 Cf Attanasio, L. (2021). Non abbiamo ancora superato i decreti sicurezza di Salvini.
    Domani. Available at www.editorialedomani.it/economia/disuguaglianze/non-abbiamo-
    ancora-superato-i-decreti-sicurezza-di-salvini-fdivyf0u [accessed 7 October 2021].

18 Eckersley also makes this point in (Eckersley, 2015)

19 For a detailed description, see (Kent & Behrman, 2018, p. 152-ss).

20 Republic of the Maldives Ministry of Environment, Energy and Water, Report on the
    First Meeting on Protocol on Environmental Refugees: Recognition of Environmental
    Refugees in the 1951 Convention and 1967 Protocol Relating to the Status of Refu-
    gees, Male, Maldives, 14–15 August 2006, cited in (Biermann & Boas, 2008).

21 See Global Compact on Refugees, para 8.

22 Norwegian Refugee Council/Internal Displacement Monitoring Centre (NRC/IDMC), The
    Nansen Conference: Climate Change and Displacement in the 21st Century, 7 June 2011,
    available at: www.refworld.org/docid/521485ef4.html [accessed 13 October 2021].

23 See the Peninsula Principles on Climate Displacement Within States, 18 August 2013,
    Principle 1.

24 See MacFarquhar, N. (2009, May 28). Refugees Join List of Climate-Change Issues.
    The New York Times, quoted in (Campbell, 2012).

25 The Dhaka Principles for Migration with Dignity (also known as the "Dhaka Princi-
    ples") are a set of principles aimed at promoting the protection of human rights migrant
    workers throughout the recruitment process until their work contracts end. The Dhaka
    Principles targeted all industry sectors worldwide and are available at www.ihrb.org/
    dhaka-principles [accessed 18 October 2021].

26 See Caramel, L. (2014, July 1). Besieged by the rising tides of climate change, Kiri-
    bati buys land in Fiji. *The Guardian*, available at www.theguardian.com/environ-
    ment/2014/jul/01/kiribati-climate-change-fiji-vanua-levu [accessed 18 October 2021].

Reference list

Benvenuti, M. (2019). Il dito e la luna. La protezione delle esigenze di carattere umanitario
    degli stranieri prima e dopo il decreto Salvini. *Diritto, Immigrazione e Cittadinanza*,
    1(1), 1–39.

Biermann, F. (2018). Global governance to protect future climate refugees. In S. Behr-
    man & A. Kent (Eds.), *Climate refugees: Beyond the legal impasse?* (pp. 265–277). New
    York: Routledge. https://doi.org/10.4324/9781315109619

Biermann, F., & Boas, I. (2008). Protecting climate refugees: The case for a global proto-
    col. *Environment*, *50*(6), 8–17. https://doi.org/10.3200/ENVT.50.6.8-17

Biermann, F., & Boas, I. (2010). Preparing for a warmer world: Towards a global govern-
    ance system to protect climate refugees. *Global Environmental Politics*, *10*(1), 60–88.
    https://doi.org/10.1162/glep.2010.10.1.60

Borras, S. (2016). Migration with dignity for climate justice. In A. Westra, L., Gray, J., &
    D'Aloia (Ed.), *The common good and ecological integrity: Human rights and the sup-
    port of life* (pp. 239–254). Routledge.

Campbell, J. (2012). Climate-induced community relocation in the pacific: The mean-
    ing and importance of land. In J. McAdam (Ed.), *Climate change and displacement:
    Multidisciplinary perspectives* (pp. 57–79). Oxford: Hart Publishing. https://doi.
    org/10.5040/9781472565211.ch-004

Cassegård, C., & Thörn, H. (2018). Toward a postapocalyptic environmentalism? Responses to loss and visions of the future in climate activism. *Environment and Planning E: Nature and Space, 1*(4), 561–578. https://doi.org/10.1177/2514848618793331

Clement, V., Rigaud, K. K., de Sherbinin, A., Jones, B., Adamo, S., Schewe, J., . . . Shabahat, E. (2021). *Groundswell part 2: Acting on internal climate migration.* Washington, DC: The World Bank.

Connell, J. (2003). Losing ground? Tuvalu, the greenhouse effect and the garbage can. *Asia Pacific Viewpoint, 44*(2), 89–107.

Cooper, J. B. (1997). Environmental refugees: Meeting the requirements of the refugee definition. *New York University Environmental Law Journal, 6*(2), 480–529.

Di Pietro, F. (2021). La nuova disciplina dei permessi per calamità, per cure mediche, per motivi religiosi, per attività sportiva, per lavoro artistico, per ricerca. In M. Giovannetti & N. Zorzella (Eds.), *Immigrazione, protezione internazionale e misure penali* (pp. 73–75). Pisa: Pacini Giuridica.

Docherty, B., & Giannini, T. (2009). Confronting a rising tide: A proposal for a convention on climate change refugees. *Harvard Environmental Law Review, 33*(2), 349–403.

Eckersley, R. (2015). The common but differentiated responsibilities of states to assist and receive 'climate refugees.' *European Journal of Political Theory, 14*(4), 481–500. https://doi.org/10.1177/1474885115584830

Ferracioli, L. (2014). The appeal and danger of a new refugee convention. *Social Theory and Practice, 40*(1), 123.

Hassine, K. (2019). *Handling climate displacement.* Cambridge: Cambridge University Press. https://doi.org/10.1017/9781316999264

Hirsch, E. (2015). "It won't be any good to have democracy if we don't have a country": Climate change and the politics of synecdoche in the Maldives. *Global Environmental Change, 35*, 190–198. https://doi.org/10.1016/j.gloenvcha.2015.09.008

Hodgkinson, D., Burton, T., Anderson, H., & Young, L. (2010). The hour when the ship comes in: A convention for persons displaced by climate change. *Monash University Law Review, 36*(1), 69–120.

Hulme, M. (2010, September). Where next for the IPCC. *Research Fortnight, 8.*

Hush, E. (2017, September 7). Developing a European model of international protection for environmentally-displaced persons: Lessons from Finland and Sweden. *Columbia Journal of European Law (Prelim. Ref. Blog.).* Retrieved from http://blogs2.law.columbia.edu/cjel/2017/09/07/

IDMC. (2021). *Global report on internal displacement.* Geneva: IDMC.

Juthe, A. (2016). Classifications of arguments by analogy part I. A comprehensive review of proposals for classifying arguments by analogy. *Cogency, 8*(2), 51–99.

Kälin, W. (2010). Conceptualising climate-induced displacement. In J. McAdam (Ed.), *Climate change and displacement: Multidisciplinary perspectives* (pp. 81–104). Oxford: Hart Publishing.

Kent, A., & Behrman, S. (2018). *Facilitating the resettlement and rights of climate refugees: An argument for developing existing principles and practices.* Abingdon: Routledge. https://doi.org/10.4324/9781351175708

Klepp, S., & Herbeck, J. (2016). The politics of environmental migration and climate justice in the pacific region. *Journal of Human Rights and the Environment, 7*(1), 54–73. https://doi.org/10.4337/jhre.2016.01.03

Kupferberg, J. S. (2021). Migration and dignity – relocation and adaptation in the face of climate change displacement in the pacific – a human rights perspective. *The International Journal of Human Rights,* 1–26.

Mayer, B. (2016). *The concept of climate migration: Advocacy and its prospects*. https://doi.org/10.4337/9781786431738

Mayer, B. (2018). Who are "climate refugees"? Academic engagement in the post-truth era. In S. Behrman & A. Kent (Eds.), *Climate refugees: Beyond the legal impasse?* (pp. 89–100). Faculty of Law, Chinese University of Hong Kong: Taylor and Francis. https://doi.org/10.4324/9781315109619

McAdam, J. (2011). Swimming against the tide: Why a climate change displacement treaty is not the answer. *International Journal of Refugee Law, 23*(1), 2–27.

McAdam, J. (2012). *Climate change, forced migration, and international law*. Oxford University Press. https://doi.org/10.1093/acprof:oso/9780199587087.001.0001

McAdam, J. (2020). Protecting people displaced by the impacts of climate change: The UN human rights committee and the principle of non-refoulement. *American Journal of International Law, 114*(4), 708–725. https://doi.org/DOI: 10.1017/ajil.2020.31

Nixon, R. (2011). *Slow violence and the environmentalism of the poor*. Cambridge: Harvard University Press. https://doi.org/10.4159/harvard.9780674061194

Pèrez, B. F. (2018). Beyond the shortcomings of international law: A proposal for the legal protection of climate migrants. In *Climate refugees: Beyond the legal impasse?* (pp. 214–229). Abingdon: Routledge. https://doi.org/10.4324/9781315109619

Platform on Disaster Displacement. (2018a). State-led, regional, consultative processes: Opportunities to develop legal frameworks on disaster displacement. In *'Climate refugees': Beyond the legal impasse?* (pp. 126–154). Abingdon: Routledge.

The Platform on Disaster Displacement. (2018b). Platform on disaster displacement, follow-up to the Nansen initiative: Addressing the protection needs of persons displaced across borders in the context of disasters and climate change. In R. McLeman & F. Gemenne (Eds.), *Routledge handbook of environmental displacement and migration* (pp. 421–425). Abingdon, Oxon: Routledge.

Prieur, M. (2018). Towards an international legal status of environmentally displaced persons. In *Climate refugees: Beyond the legal impasse?* (pp. 233–259). Abingdon: Routledge. https://doi.org/10.4324/9781315109619

Prieur, M., Marguénaud, J.-P., Monédiaire, G., Bétaille, J., Drobenko, B., Gouguet, J.-J., . . . Shelton, D. (2008). Draft convention on the international status of environmentally- displaced persons. *Revue Européenne de Droit de l'Environnement, 12*(4), 395–406. https://doi.org/10.3406/reden.2008.2058

Scissa, C. (2021). La protezione per calamità: Una breve ricostruzione dal 1996 ad oggi. *Forum Di Quaderni Costituzionali, 1*, 136–147.

Shacknove, A. E. (2010). Who is a refugee? In H. Lambert (Ed.), *International refugee law* (pp. 163–173). Farnham, Surrey England: Ashgate.

United Nations Development Programme. (2016). *Tulele peisa, papua new guinea. Equator initiative case study series*. New York, NY: United Nations Development Programme. https://www.equatorinitiative.org/wp-content/uploads/2017/05/case_1473429470.pdf

Walker, B. (2017). *An Island nation turns away from climate migration, despite rising seas*. Retrieved October 18, 2021, from https://insideclimatenews.org/news/20112017/kiribati-climate-change-refugees-migration-pacific-islands-sea-level-rise-coconuts-tourism/

Williams, A. (2008). Turning the tide: Recognizing climate change refugees in international law. *Law and Policy, 30*(4), 502–529. https://doi.org/10.1111/j.1467-9930.2008.00290.x

4   The justice dilemma –
"Climate Refugees" as a case
of environmental (in)justice

The lack of legal status of "climate refugees" is not only a technical issue of filling a legal gap but also, and above all, a question of justice. People fleeing environmental disruptions can either be treated unlike similar/same cases, suffer disproportionate discriminatory effects deriving from climate-related events, or receive less if no assistance to cope with exacerbated vulnerabilities and inequalities. In other words, shortcomings in international law are becoming a factor of vulnerability in itself able to jeopardize the human rights of the so-called climate refugees.

Tragically, people on the move in the context of environmental disruptions are likely to face three scenarios of injustice.

The first scenario occurs when they become internally displaced within their own countries. As mentioned in previous chapters, the Guiding Principles on Internal Displacement (UNHCR, 1998) define internally displaced persons (IDPs) as

> persons or groups of persons who have been forced or obliged to flee or to leave their homes or places of habitual residence, in particular as a result of or in order to avoid the effects of armed conflict, situations of generalized violence, violations of human rights or natural or human-made disasters, and who have not crossed an internationally recognized State border.
>
> (Guiding Principles on Internal Displacement, 1998)

Despite the high numbers of IDPs by natural disasters (30.7 million) (IDMC, 2021), the guiding principles on internal displacement is still the only non-binding, soft-law protection framework provided for by international law. The only example of a treaty giving these non-binding principles the force of law is the 2009 African Union Convention for the Protection and Assistance of Internally Displaced Persons in Africa (Kampala Convention) adopted by African states in October 2009.[1] In force only in Africa, it includes an explicit obligation "to protect and assist persons who have been internally displaced due to natural or human made disasters, including climate change."[2] This weakness in international law implies that the same/similar cases are treated unlike. In Africa, the protection and assistance of internally displaced people due to environmental disruptions largely vary depending upon the African country where the internal movement occurs. As mentioned, some African states have already ratified the Kampala Convention,

DOI: 10.4324/9781003102632-4

but some have not ratified it yet. As a result, the same/similar cases are treated differently based on the African State where the movement occurs. Similarly, when those people move internally within countries other than the African States, their protection and assistance largely depend upon the willingness of States to implement the 1998 Guiding Principles on Internal Displacement, which is, indeed, a soft-law instrument.

In practical terms, such legal shortcomings raise issues of formal justice in the first place as similar/same cases are treated unlike. Indeed, this outcome infringes the principle of formal justice, establishing that if two cases are relevantly similar, they ought to be treated alike (Juthe, 2016).

In the second place, internal displacement results in overcrowding, competition over scarce resources, including scarce water and food supplies, and increased conflict exposure. These consequences exacerbate existing inequalities, thus producing a discriminatory effect against the most vulnerable countries, communities, and individuals. Ultimately, these inequalities depend upon legal shortcomings that constitute a factor of vulnerability in itself.

The second scenario entails that displaced people cross borders. In most cases, they are more likely to move to neighbor developing states. The choice is motivated mainly by geographic proximity as those countries are easier to reach, allow better control over things left behind, and return when conditions permit. However, neighbor developing states are often also affected by climate change and usually share the same vulnerabilities. It follows that mass migration in the aftermath of environmental disruptions may negatively impact those neighboring states, thus increasing the environmental deterioration, especially in the areas receiving migrant populations. The issue is well-known and was largely discussed during a Symposium of 60 international experts examining links between mass migration and environmental deterioration. The meeting was held at Chavannes-de-Bogis near Geneva between 21 and 24 April 1996. The Conference Report particularly emphasized the most common environmental impacts associated with mass migration:

> deforestation, soil erosion, and water contamination or depletion. In turn, those impacts result in a reduction in migrants' quality of life, the depletion of natural resources crucial to the local economy, economic and political challenges in the areas receiving migrant populations, and difficulties in supporting sustainable development in these areas (International Organization for Migration, 1996). In other words, mass migration can overstress the carrying capacity of neighbor developing states, thus increasing inequality across countries and jeopardizing the living conditions of host communities and migrants. This scenario raises issues of distributive justice between States that have been only partially addressed by introducing the principle of burden- and responsibility-sharing in the 2018 Global Compact on Refugees (GCR). As explained in Chapter 2, the GCR is a soft-law instrument that by nature cannot ensure its effective implementation. Again, the weakness of

the international legal framework and international governance increases inequalities and does not ensure the same degree of protection of human rights from environmental risks.

The third scenario is that people fleeing environmental disruptions move to developed countries, thus mostly ending up in immigration detention facilities as irregular migrants.

Mostly sited in environmentally fragile areas, these settlements were initially conceived as emergency camps for providing temporary shelter to displaced people. However, it is estimated that the average lifespan of such settlements is up to 17 years (Moore, 2017), where people experience protracted detention before resettlement or eventually repatriation. Protracted detention implies restrictions in terms of freedom of movement, education, legal employment access, political participation, and access to justice. People in such situations totally depend on international assistance to meet their basic needs and are substantially unable to live what Sen has defined as the life they have reason to value (Sen, 1999). Further, they are more often subjected to sexual and physical violence committed by staff at the facilities. Examples include the Australian-run camps on Nauru and PNG's Manus Island, where the office of ICC prosecutor denounced dangerous and harsh conditions up to the form of crime against humanity and torture.[3] Similar to the second scenario, the disproportionately high population densities in immigration detention facilities create tensions between asylum seekers and host communities. Anti-immigrant parties often portray asylum seekers as economic migrants that can threaten the host society's prosperity and security, thus fueling frictions between host communities and asylum seekers that prevent their integration and full consideration as members of the host communities.

When looking at the lack of recognition of climate refugees as new legal subjectivities, this focus on justice allows to re-position the debate around human beings and their human rights. The next paragraphs will examine the mechanisms of objectification that have jettisoned these aspects from the debate around the figure of "climate refugees" through the lens of decolonial environmental justice. In particular, I will dwell upon what decolonial environmental justice is, why it matters for this debate, and what contributions can bring to the search for recognition of "climate refugees."

What (decolonial) environmental justice is and why it matters for "Climate Refugees"

What precisely is the "injustice" where people forced to migrate because of environmental disruptions are concerned? The three scenarios described earlier have shown that shortcomings in the international legal framework and international governance ultimately prevent ensuring the same/sufficient degree of protection from environmental risks triggering migration. Such an unfair outcome can be well-conceived as a form of *environmental injustice*. By this term, I refer to the unequal power relations generating the disproportionate sharing of environmental risks and goods at the expense of marginalized, low-income, discriminated communities. (Pellow, 2017)

The core idea of environmental justice is that all people should be treated equally with respect to the development, implementation, and enforcement of environmental laws, regulations, and policies.

In practical terms, all people should be recognized as political actors having a meaningful involvement in the decision-making process related to environmental issues.[4]

The term environmental justice can be traced back to the 70s when early mobilizations in the US (McGurty, 2000; Phillips, Hung, & Bosela, 2007) claimed the equal protection of low-income and ethnic minority communities from toxic waste dumping. First studies (Bullard, 1990; United Church of Christ Commission for Racial Justice, 1987; the United States General Accounting Office, 1983) demonstrated that race was the main factor in determining the siting of (toxic) waste facilities. As the most affected communities in the US context were African Americans and ethnic minorities, this phenomenon was initially termed *environmental racism* (Bullard, 1990; Pulido, 1996; Taylor, 2000).

As noted by Pellow, such a first generation of EJ studies were more concerned with documenting environmental injustices, whereas the second changed its trajectory toward a greater interdisciplinary engagement and further improvement over the methodologies and epistemologies (Pellow, 2016). This new direction in EJ scholarship was labeled by Pellow and Brulle *Critical Environmental Justice Studies* (Pellow & Brulle, 2005). One of its main merits has been shifting the focus of environmental justice beyond the sole dimension of distribution toward greater emphasis on recognition. In particular, it expanded the spectrum of *misrecognition* beyond race/income by including gender, sexual orientation, and the non-human world.

Among the environmental justice theorists, Schlosberg's contribution to building up a consistent analytical framework is one of the most popular. According to the author, what makes *environmental* justice peculiar is its four dimensions: distribution, recognition, participation, and capabilities (Schlosberg, 2007). The unequal distribution of environmental risks and goods is strictly related to the *misrecognition* of the most disadvantaged groups within societies. Importantly, marginalized groups have fewer financial and cultural resources to resist structural injustices/externalities produced by the capitalist system. Their voices are usually not heard, silenced, and often ignored by policy-makers throughout the decision-making process. Thus, the lack of recognition does prejudice the meaningful *participation* of all groups in the political arena. Ultimately, policies, regulations, and laws disregard the needs and claims of the most disadvantaged groups, thus creating *sacrifice zones* where the resident community's well-being (understood as a person's human *capabilities*) is seriously jeopardized.

Over the last decade, the "trespassing" of environmental justice outside the US has increased its nuances due to its utilization in several contexts worldwide

(Holifield, Chakraborty, & Walker, 2017). In this regard, a robust corpus of literature has shown that environmental justice has found different applications not only in its birthplace, the US (Bullard, 1990, 1993, 1994), but also in Europe (Agyeman & Evans, 2004; Laurent, 2011), Latin America (Carruthers, 2008),[5] Asia (Basu, 2018; Fan & Chou, 2018), Africa (Ssebunya, Morgan, & Okyere-Manu, 2019), Oceania (Schlosberg, Rickards, & Byrne, 2018), and the Arctic region (Shaw, 2018).

As a result, defining EJ in uniform terms has become challenging, with different ways of thinking this concept weakening the emergence of a uniform conceptual framework up to what Benford has termed the "frame-over extension": extending the EJ frame so broadly can lead to the point that it loses much of its original mobilizing power (Benford, 2005, p. 47).

In this view, reflections from Latin American scholars have pushed this debate even further with a re-engagement with (environmental) racism, colonialism, and forms of dehumanization for rethinking the entire EJ paradigm toward a "decolonial turn." This expression refers to the need to decolonize environmental justice in the face of persistent colonial values (i.e., coloniality) within the EJ paradigm (Álvarez & Coolsaet, 2020; Maldonado-Torres, 2011; Rodriguez, 2020).

The core idea is that colonialism did not end between 1945 and 1960 with the rise of new states in Asia and Africa becoming independent by their European colonial rulers, but survives in the form of coloniality: the set of practices resulting from the unequal power relations inherited by colonialism still at work within contemporary societies (Álvarez & Coolsaet, 2020; Maldonado-Torres, 2011). As for the EJ paradigm, the coloniality of justice reported by this strand of research emerges from the use of Western categories to define *justice* and *environment*, the overemphasis on distribution dimension, and the overly state-centric approach (Álvarez & Coolsaet, 2020).

Developed under the banner of Decolonial Environmental Justice, this frontier of EJ is rooted in the concept of coloniality conceived in its threefold understanding of coloniality of power, knowledge, and being (Álvarez & Coolsaet, 2020; Escobar, 2007; Grosfoguel, 2007; Mignolo, 2012; Quijano, 2000; Rodriguez, 2020).

So far, the coloniality of power has been examined along two axes: (1) the racial difference between Europeans and non-Europeans, with colonial/modern capitalism drawing economic, cultural, ontological, and epistemological forms of subjugation; and (2) the use of Western institutional forms of power, such as the nation-state, in non-Western societies where different cultures and legal pluralism exist (Quijano, 2000). Imposed top-down, it consists of a form of exploitation and economic and political domination based on race and the international division of labor and political power. In Quijano's words,

> That specific basic element of the new pattern of world power that was based on the idea of 'race' and in the 'racial' social classification of world population – expressed in the 'racial' distribution of work, in the imposition of new 'racial' geocultural identities, in the concentration of the control of productive

resources and capital, as social relations, including salary, as a privilege of 'Whiteness' – is what basically is referred to in the category of power.

(Quijano, 2000, p. 218)

Similarly, the coloniality of knowledge emerges from the distinction between the (superior) European knowledge and (inferior) non-European knowledges. According to this hierarchy, only the European knowledge is considered scientifically valid, theoretically sound, detached from geo-historical conditions and, therefore, with universal validity. By contrast, non-European knowledges and worldviews are conceived as too local, traditional, and ultimately too place-based and context-related to be applied globally. As a result, this type of coloniality disqualifies all other knowledges and their theoretical relevance, thus perpetuating an epistemological form of subjugation.

Finally, the coloniality of being has to do with the effects of coloniality in the lived experience of colonized. It consists of constructing hierarchical subjectivities with geographical and epistemic borders separating those "normal" and "superior" from others "inferior" and "unworthy" (Álvarez & Coolsaet, 2020; Maldonado-Torres, 2007). This type of coloniality differs from the other two as it is not imposed top-down. Nevertheless, it seriously affects the self-image of the colonized and the perception of their world. In particular, it creates what Fanon called *zones of non-being* (Fanon, 1967), thus dehumanizing subjects and communities.

The following paragraphs will employ the Decolonial Environmental Justice by using this tripartite notion of coloniality to uncover, and try to overcome, the coloniality that prevents the recognition of climate refugees as new legal subjectivities.

The Threefold Injustice of "Climate Refugees": Coloniality of Power, Knowledge, and Being

The coloniality of power

If we look at the figure of "climate refugees" through the lens of Decolonial Environmental Justice, we can easily identify the three forms of coloniality described earlier.

As for the coloniality of power, both axes (racial difference/use of Western institutional forms of power) emerge from the colonial/modern capitalism influencing mobility and categories of institutional power on the basis of race and the international division of labor.

As pointed out by Baldwin, the presence of racial power in climate-change and migration discourse can be found in three specific racial tropes: naturalization; the loss of political status; and ambiguity (Baldwin, 2013, p. 1486). By figuring the "climate refugee" as both threat and victim,[6] the Western narrative portrays a racial Other with no room for agency and no space for his history. The major force compelling the "climate refugee" to flee is the *external nature*, the biophysical world itself. Thus, the trope of naturalization means silencing or ignoring other *political* forces that cause human mobility, such as physical violence, dispossession, or discrimination (Baldwin, 2013, p. 1480). The loss of state territory and transboundary migration constitutes the trope of the loss of political status. In both cases, people *without rights* may become *a kind of noncitizen out of place in the Westphalian order of nation-states* (Baldwin, 2013, p. 1482). Finally, the trope of ambiguity has to do with the fuzziness of "climate refugee." Lacking a conceptual and legal definition, this racial Other remains undefined, not calculable, and ultimately not fully represented and/or recognized.

Such a racialization of the figure of "climate refugee" results in a violent form of environmental racism where the most affected are those whose lifestyle entirely depends upon the specific territory they occupy: indigenous Arctic people, Inuit, and Blacks fleeing from the droughts of Northern Africa (Westra, 2009). Not only anthropogenic climate change adversely impacts Indigenous peoples' material well-being[7] and cultural ecosystem services necessary for their social reproduction,[8] but it ultimately results in what Whyte termed a colonial déjà vu (Whyte, 2017). With this term, Whyte refers to the fact that climate change is more likely to re-establish colonial mobility patterns aimed at restricting indigenous peoples' mobility. In this view, climate change impacts do not create new forms of mobility injustices but rather exacerbate and interact with past injustices concerning Indigenous mobility's restriction through treaty violations, land seizure, forced removal, and reservations (Whyte, Talley, & Gibson, 2019).

The history of unfair mechanisms aimed at generating and maintaining forms of mobility injustice can be traced back to the Cold War. At that time, "those escaping from communist countries were often welcomed (in the US, Canada, Australia, or Western Europe), while refugees from colonial liberation wars in Africa and Asia generally ended up in camps in these regions with little hope of resettlement" (Castles, 2006b, p. 19).

To date, the same pattern aimed at limiting the mobility of non-Europeans evolves out of the general trend to deal with migration through hardline anti-immigration policies. Indeed, measures taken by the majority of the developed countries aim at curbing migratory flows by preventing landings and restricting the opportunities of asylum seekers (Kukathas, 2016).

The racial difference between white, European migrants allowed to move and non-white, non-European, migrants even prevented land emerges clearly by the free-border Schengen Area (Fröhlich, 2017). Created in 1995, it is considered the world's largest visa-free zone. Free movement of persons, however, enables only EU citizens (with the sole exception of non-EU nationals living in the EU or visiting the EU as tourists, exchange students, or for business purposes) to travel, live, and work in an EU country without being subjected to border checks or particular formalities.[9]

As a result, the side effects of the Schengen Area has been the creation of a "Fortress Europe" where the EU's border agency, Frontex, is tasked with defending Europe from the "invasion" of migrants by fences, warships, and a surveillance system of satellites and drones. By examining this militarized border making fueled by security narratives (Bettini, 2013), Parenti coined the term "armed lifeboat," referring to rich countries that use their privilege to protect their own elites while shutting out "climate refugees"(Parenti, 2011). As also observed by Park and Pellow, such a militarized border making based on race/fear of the Other results in the form of environmental privilege, having the unfair outcome of securing environmental amenities to the wealthier white-European communities at the expense of the environment and wellbeing of the non-white, non-European, poorest communities (Park & Pellow, 2019).

The coloniality of power also reaffirms this racial difference between Europeans and non-Europeans along the line of the international division of labor based on globalized socio-economic inequalities. Supported by the more recent narrative of "migration as adaptation," migration policies are designed to facilitate high-skilled migrants who successfully adapt to the labor market and the society of destination countries while restricting access to low-skilled migrants failing to pass the "resilience test."

As pointed out by Bettini,

> Ecologically vulnerable populations are to be transformed into adaptive subjects, hoping to reach a dynamic form of socio-economic development based on the idea of resilience goals (Agrawal & Lemos, 2015). States become enforcers and facilitators rather than active actors, with the market and self-organisation becoming key mechanisms.
>
> (Bettini, Nash, & Gioli, 2017, p. 7)

This neoliberal form of migration management (Bettini, 2014) has materialized through the high-skilled immigration policies promoted by the EU over the last decades (Kahanec & Zimmermann, 2012). Such policies are justified due to skill mismatch existing in the EU market labor and the positive impacts skilled migrants may bring about. Evidence shows that the EU needs international migrants' skills to cope with aging populations, demographic deficit, stalled economic growth, lack of innovation potential, and skilled workforces (Foresight, 2011, p. 183; Kahanec & Zimmermann, 2012). In this context, people "knocking on western doors" are implicitly selected due to their expendability in the labor market. In practical terms, they are welcome as long as/until they are useful to host countries' economies.

Examples of these temporary/circular migration schemes[10] include seasonal/guest-worker programs, yearly labor migration quota, foreign "contract workers," and education programs[11] (see Chapter 3 of this volume). Such circular migration schemes have allowed the entry of limited numbers of high-skilled workers to meet specific labor needs in some EU countries such as Germany, Netherlands, Norway, Ireland, Belgium, Sweden, Greece, Italy, and Spain (Castles, 2006a, p. 14). What they all have in common, however, is that they do not necessarily lead to permanent settlement.

> The Global Commission on International Migration saw important development opportunities arising from more fluid circular migration, with people able to move

back and forth more easily because 'the old paradigm of permanent migrant settlement is progressively giving way to temporary and circular migration'.

(Foresight, 2011, p. 184)

However, what happens to those with a less economic appeal, such as lower-skilled migrants, women, children, disabled, and those who are less educated, poorer, and require more health assistance? Such policies are more likely to discriminate against the most vulnerable groups unless they transform into adaptive, resilient subjects.

In this regard, Germany is a case in point. Indeed, its 2004 Zuwanderungsgesetz (immigration law) indicates two different pathways for high-skilled and lower-skilled workers, respectively. High-skilled workers can apply immediately for permanent residence and have facilitated entry into the destination countries. By contrast, lower-skilled workers can only come in through small-scale temporary worker schemes for specific sectors or undocumented migrants (Castles, 2006a).

Working conditions are often established through bilateral agreements with countries of origin, thus remaining a few numbers with limited possibility of scaling up. Further, those imply that the recruitment of non-EU workers is possible only if no EU workers are available (Castles, 2006a).

Unsurprisingly, these temporary/circular migration schemes have been proposed for further application to "climate migration" (Felli, 2013, p. 353; Foresight, 2011, pp. 184–185).

Above all, in times of climate stress (e.g., drought, floods, extreme weather events), people affected may opt for an adaptive migration. This latter often implies sending a male family member elsewhere to find paid labor until environmental conditions enable him to return home. The aim is to help top-up a family's income through remittances, thus reducing the draw on local resources (Kälin & Schrepfer, 2012). Although these measures can increase the resilience of those who stay at home during the climate crisis (e.g., by remittances), they are likely to deprive such circular migrants of the most basic rights in destination countries (Felli, 2013). The precariousness of their jobs justifies reduced entitlement to basic rights (e.g., unemployment benefits, supplementary pension, residence permits), not least because these "second-class" workers do not have any political (citizenship) rights (Felli, 2013). Ultimately, the limit of such neoliberal migration management is that the resilience goals are achieved through individual behaviors fostered by the labor market rather than collective political action. This outcome has discriminatory effects against vulnerable groups.

Further, the coloniality of power materializes in categories of institutional power, like the nation-state, that does not fit environmental issues, such as climate migration. The centrality of nation-states does not capture the changing nature of today's refugeehood. Western institutional categories, such as the nation-state, do not always address the context-specific forms of power existing in non-Western countries. Notably, state sovereignty is at stake because of armed conflicts, but more often because of the plural legal systems prevalent in non-Western countries (see Chapter 2).

Finally, the centrality of nation-states does not capture the empirical reality of today's human mobility, where natural disasters triggered over three times the number of displacements resulting from armed conflicts in 2020 (IDMC, 2021). This reversal of the traditional migration patterns suggests that the use of nation-states is ineffective in environmental disruptions, which have impacts far beyond the territorial and socio-cultural limits of the single countries.

On top of that, this migration pattern change is more likely to question the logic of culpability. This latter underpins traditional attempts to theorize refugee flows over the role of states in "making" refugees following wars or armed conflicts (Piguet, 2019). In this state-centric perspective, states are "culpable" for having "produced refugees" as in the time of war, they cannot ensure their citizens' basic rights, thus creating a "refugee" problem to be solved by the international community. By contrast, climate migrations ontologically differ from this pattern as their movement is either triggered by natural disasters or anthropogenic climate change (e.g., CO_2 emissions) caused mainly by polluting companies, which are non-state actors.

The coloniality of knowledge

The superiority of the Eurocentric knowledge regime manifests in the dogmatic 1951 Refugee Convention's definition of a refugee and its overly narrow interpretation. As examined in Chapter 2, even the 2018 Global Compacts reaffirm the centrality of this definition by making clear that only refugees as defined by the 1951 Convention Relating to the Status of Refugees and its 1967 Protocol are entitled to the specific international protection.

In doing so, the 2018 Global Compacts mark (once again) a hierarchy of knowledges where only the Eurocentric 1951 Refugee Convention has universal validity regardless of the changed geopolitical and geohistorical conditions.

The resistance to extending the definition of a refugee to cover "climate refugee," either by a new legal instrument or by a broader interpretation, stems from this overemphasis on nation-states as categories of power and the persistence of a geopolitically outdated reading of this subject matter.

Above all, the centrality of nation-states survives in the 2018 Global Compact on Refugees (GCR), imbued with references to the Westphalian principle of the equal right to sovereignty and the primacy of national policies and priorities in governing large-scale influx of refugees (see Chapter 2).

Further, the insistence over the Refugee Convention definition and its emphasis on persecution as an essential criterium of refugeehood seems still rooted in the strategic conceptualization given by the Western States to the Refugee Convention during the Cold War. In that geopolitical context, the five grounds of persecution had to do with those human rights whose protection was not ensured in the eastern bloc (Hathaway & Foster, 2015). Hathaway noted that this strategy prioritized the protection of those whose flight was motivated by pro-Western political values.

The Cold War is over, yet the "cold" wording of the 1951 refugee definition seems to prevail over doctrinal and judicial case law developments and the current geopolitical context. Above all, the lexicon of the Cold War survives in its reno-vated form aimed at protecting Western liberal democracy, and the neoliberal pro-ject ultimately based on the value of the free market.

Not surprisingly, such a neoliberal agenda is pursued by shaping the termi-nology over terms like "migrants" or "displaced people" while disqualifying the most problematic category of a "refugee' (see Chapter 1). In doing so, states can easily reaffirm their discretionary control over who is in and out of their physical and epistemic borders while disregarding the legal obligations that would come up if they employed the refugee category for people on the move in the context of climate change. To date, such a persistent centrality of nation-states is signifi-cantly percolating the debate on "climate refugees" through discourses advanced by states by the Platform on Disaster Displacement (PDD): one of the most influ-ential actors in the field within the cross-governance system of the Task Force on Displacement (see Chapter 2).

The coloniality of being

The overly state-centric approach when dealing with climate refugees leads to the construction of hierarchical subjectivities based on the national belonging and expendability in destination countries' market labor. These mechanisms of raciali-zation are likely to attribute different (or any) rights according to the selected category, thus resulting in discrimination, reification, and dehumanization of sub-jects and communities.

Paradoxically, such emerging coloniality of being is partially caused (and defi-nitely not solved) by the so-called EJ mainstream. Crucially, scholars have often advocated environmental justice to both re-politicize the issue of "climate refu-gees" and to shed light on the responsibilities of states and the international com-munity. Above all, EJ has mainly been invoked to reframe the debate by focusing on the root causes of human migration.

The EJ syllogism, still widely contested in its premises, was outlined as follows.

- Major Premise: Climate change is caused by human activities.
- Minor Premise: Climate change's impacts trigger human mobility.
- Conclusion: Therefore, human mobility will decrease if human activities stop causing climate change.

It follows that EJ advocates demand more States' actions in terms of climate change mitigation.

However, the focus on states' responsibilities to assist and receive "climate refugees" has adverse side effects. The main reason is that states disregard their responsibilities because this syllogism is valid but false. The two premises are not supported by sufficient evidence, so that the conclusion is false. Droughts, floods, extreme weather events may also be nature-made disasters as well as human mobility is a multi-causal phenomenon where environmental degradation is only one among many other factors.

In this view, rather than influencing States to adopt an EJ perspective, there is a danger that EJ is pushed into a state-centric approach. As outlined in the introduction, the original aim of EJ was to prevent environmental risks from being produced in the first place and distribute them equally when produced. By contrast, if a state-centric approach integrates EJ, its original aim risks being misinterpreted, thus leading to (or even justifying) the following paradoxical results.

The first side result is that "climate refugees" be treated as undesirable (and even illegal), material flows likely to trigger environmental conflicts (e.g., frictions with host communities). To the same extent as waste, "climate refugees" risk to be framed as a problem to be prevented at source, useful material flows if reusable/recyclable in the labor market of destination countries, or unwanted, material flows to be disposed of when not reusable/recyclable.

The prevention of "climate refugees" at source may imply measures for deterring the landings to preventing people on the move from arriving on the territory of host countries in the first place (e.g., close ports to landings, repel people trying to cross the border, etc.).[12] Secondly, it may require strengthening investments in developing countries to keep people in their countries of origin.

Finally, and above all, it would question the very existence of "climate refugees" by rejecting the term and any studies suggesting its use.

Notably, the prevention of "climate refugees" at source is, above all, epistemic. It occurs by avoiding providing a univocal definition, denying "climate refugees" a legal recognition, and ultimately generating ambiguity. In other words, their existence is prevented at source by impeding the formation of a specific, distinct subjectivity justifying a definition and a legal status, thus hindering them become subjects with rights and duties.

When people on the move due to environmental disruptions succeed in reaching developed countries, they mostly end up in immigration detention facilities as *irregular migrants*. Treated as unworthy, illegal, unwanted, *disposable migrants*, they are disposed of in *social landfills* as inanimate objects. Therefore, those people are dispossessed, confined in *zones of non-being* (Fanon, 1967), and robbed of their status as human beings/(legal) subjects.

> Yet dispossession is precisely what happens when populations lose their land, their citizenship, their means of livelihood, and become subject to military and legal violence.
>
> (Butler & Athanasiou, 2015, p. 3)

Thus, they are dehumanized and have no access to international protection, healthcare, education, work, and risk being sent back to their home countries while experiencing protracted refugee-like situations before resettlement. Ultimately, they are treated as objects rather than subjects with rights and duties.

Such dehumanization leads us to the *second side effect* of having an EJ state-centric approach. It implies that "climate refugees" are considered threats/burdens/objects to be equally distributed *within* the State territory, once entered State of destination, or *among* States when ended up in immigration detention camps at the external borders, for example, of the EU.

In this view, EJ risks being invoked to justify state claims concerning labeling and the allocation of "climate refugees" per quota as they were negative externalities. This danger relies on the fact that principles of triage might be easily camouflaged with principles of distributive justice. Indeed, both triage and distributive justice principles are often utilized to determine how (survival) goods ought to be distributed in conditions of scarcity (McKinnon, 2012). However, there is a substantial difference between the two.

Distributive justice principles provide moral guidance to distribute the benefits and burdens of economic activity within the society. Those include criteria to establish what is to be distributed, to whom, and the principles governing this distribution. By contrast, principles of triage imply the classification and prioritization of people in the name of efficiency. Indeed, the term *triage* stems from the medical metaphor used for sorting victims based on the "gravity of damages in order to provide (medical) support to those most in need first" (Perchinig, 2017, p. 147). For the purpose of this analysis, triage principles sort and categorize *victims* of displacement according to their economic expediency in the global (labor) market along with national belonging.

In practical terms, this logic of triage leads to this paradoxical outcome:

> triage systems categorize *people on the move* in different categories regarding *their expendability in destination countries' market's labor* in order to *get (economic) advantage to their societies* first.[13]

Justified by the conditions of scarcity,[14] such a triage logic is often (mis)represented as distributive justice by the desert principle: everyone gets what they deserve. The more skills/adaptive capacities migrants have, the more rights are expected to have. Evidence supporting this argumentation can be found in the resulting *social* hierarchy linked with the following labeling[15]:

1 *citizens* are people having full rights and access to services;
2 *refugees* are people with a special legal status as long as they meet the persecution/alienage requirements (rights to temporary protection);
3 *migrants* are conceived as commodities to be allocated in destination countries' labor markets;
4 *irregular migrants* are considered "illegal," thus they do not have access to healthcare, education, legal jobs, basic services, and risk repatriation.

In the lack of a proper categorization, people fleeing environmental disruptions might be classified as migrants, thus enjoying reduced entitlement to basic rights compared to citizens, or as irregular migrants without rights and access to basic services.

In this regard, the coloniality of being also manifests in "colonized" (self-) identification with terms they did not even choose so that their self-image is seriously affected by the lack of a specific legal status.

On the one hand, their self-image is manipulated by nationalist rhetoric, such as the "migration with dignity" motto underpinning the former relocation scheme in Kiribati (see Chapter 3). Rather than protecting their human rights, this rhetoric ultimately facilitates their reification as commodities for developed countries. On the other hand, it is determined by the legal protection they expect from tribunals or offices responsible for examining their asylum applications. Lawyers defending their rights[16] often report the lack of awareness and fear of telling their "right" story (Barca, 2014). Most people fleeing environmental disruptions omit to tell that environmental factors played a role in their movement because they are afraid of being framed as economic or (even worse) irregular migrants. As a result, their stories are often pre-constructed and distorted, with the sole purpose of being framed with the "right" category to receive protection from host countries.

Decolonizing the refugeehood

The analysis done earlier has shown how the coloniality of justice emerging from the overly narrow focus on distribution and state-centric approach ultimately results in the coloniality of power, knowledge, and being, which can jettison human beings and their rights from the debate surrounding climate-induced migration.

In particular, the overly state-centric approach risks jeopardizing both the search for recognition of climate refugees and EJ's transformative mission itself. On the one hand, the human rights of "climate refugees" are jeopardized by state inaction and its unwillingness to recognize their existence as subjects of law in the first place. On the other hand, such a state-centric approach risks having negative implications for EJ as it may fail to achieve its goal of dismantling "the mechanisms by which capital and the state disproportionately displace the social and ecological costs of production onto working-class families and oppressed peoples of color" (Faber, 1998, p. 13).

When EJ scholars claim that states take responsibility for "climate refugees," they ultimately accept state legitimacy. However, this legitimation may lead to a sort of heterogenesis of ends so that EJ claims are distorted by the state and applied as triage principles. As a result, although EJ scholars' purpose is to demand states to do justice for "climate refugees," the outcome they achieve is that EJ is substantially misinterpreted by states, which treat "climate refugees" (1) as undesirable (and even illegal), material flows likely to trigger environmental conflicts and (2) as threats to be equally distributed. Far from being environmentally just, the outcome is more coherent with principles of triage.

Ultimately, by demanding more States' actions, EJ scholars do not consider how state power is more likely to reinforce rather than correct social inequalities. As Pellow warns in an interview conducted by Laura Pulido[17]

> the state cannot be the only game in town. Yet, the movement continues to seek justice from institutions and a legal system that was never intended to offer it in the first place and may be incapable of doing so.
>
> (Pulido, 2017, p. 47)

In other words, one cannot ask for a "solution" from the subject that has created the "problem."

How to avoid such a paradoxical and unfair outcome? How can we do environmental justice for "climate refugees" while restoring environmental justice's transformative mission?

My argument is based on the following hypothesis. If the state-centric approach were replaced with a non-state-centric approach, then it would be possible to rethink the essential criterium of refugeehood while pushing the "climate refugees" governance beyond the triage governmentality.

The shift toward a non-state centric approach is possible when the following conditions are accomplished:

(1) the concept of refugeehood is decolonized;
(2) the unit of responsibility is shifted from states to non-state actors;
(3) the term responsibility is conceived as "empowerment" rather than "liability."

The first step is decolonizing the refugeehood with a broader interpretation of what is a refugee according to the current geopolitical context.

To date, a refugee is still seen from the State perspective as a problem to be solved or a burden to be equally shared. Whether this centrality of States could have made sense in the geopolitical context of the Cold War (i.e., a world having borders apparently *frozen*) does not seem valid today in a world ever more dominated by cross-border environmental conflicts.

Not only are displacements caused by natural disasters over three times those by armed conflicts (IDMC, 2021), but a growing corpus of literature has also demonstrated that armed conflicts themselves are increasingly linked by the impacts of climate change in vulnerable contexts (Ide, Brzoska, Donges, & Schleussner, 2020). In light of this, a major emphasis should be given to environmental conflicts than armed conflicts when looking at the current geopolitical context.

According to Lee, *environmental conflicts* can be defined as situations "where natural resources or natural conditions are a cause, consequence, tool, or weapon in a large-scale political dispute resolved through conflict or cooperation" (Lee, 2020, p. 40). In the current context, armed conflicts are increasingly correlated by two types of environmental disruptions: disputes related to hot areas (e.g., drought) and cold areas (e.g., melting glaciers). Lee has initially framed such an expanded geopolitical reading of environmental conflicts by distinguishing between Polar and Equatorial tension belts (Lee, 2009). In his view, there are two geopolitical features conceived as tension belts. The first is the Equatorial tension belt. It comprises South Africa and central Asia, thus largely including developing countries. Conflicts in this belt are more likely to be triggered by resource depletion and desperation. The second is the Polar tension belt, approximately located around the Polar Circles. Conflicts here are expected to be mainly driven by new resource exploitation. The changing environmental circumstances in this belt are opening new frontiers of development and resource exploitation, thus triggering frictions between local populations, who perceive this rush as a colonial déjà vu

(Whyte, 2017), and neighboring States competing for land and new opportunities (Keucheyan, 2016).

In 2020, the author provided an updated version of such a Hot Wars geopolitical context where he replaces this dichotomy between Polar and Equatorial belts with hot war and cold war clusters[18] (Lee, 2020). According to this refined version, cross-border environmental conflicts are blurring the divide between developing and developed States as they are more likely to manifest in the climate "hot-spots" (Giorgi, 2006).

In light of this, there is reason to decolonize what constitutes a refugee in the current "Hot Wars" geopolitical context while questioning the centrality of nation-states. In my view, a "decolonized" definition of a refugee is possible by a broader interpretation based on analogy by law along with the concept of vulnerability. The basic assumption is that vulnerability may serve as a *tertium comparationis* to establish an analogy between political and climate refugees regardless of the reasons why they escape from their countries of origin (e.g., persecution, war, natural disasters).

Both categories ultimately share the same vulnerabilities, being their fundamental human rights at risk upon return to their home countries.

The term vulnerability is conceived here as "a function of exposure, sensitivity to impacts and the ability or lack of ability to cope or adapt" (UNEP, 2007, p. 304).

The focus on vulnerability allows identifying, on the one hand, climate hot-spots, i.e., regions "for which potential climate change impacts on the environment or different activity sectors can be particularly pronounced" (Giorgi, 2006, p. 1). On the other hand, it helps to identify people in vulnerable situations: socioeconomically deprived communities for whom climate change impacts will urgently require humanitarian protection.

In the first case, I refer to an ecological understanding of vulnerability. In such a perspective, climate scientists can identify climate hot spots and, therefore, also the communities living there who will be hit hardest by climate change impacts.[19] In the second case, I refer to a social understanding of vulnerability. Social vulnerability deals with the different susceptibility based on social, economic, and political factors (Cutter, 2001).

This approach rooted in the concept of vulnerability might better address distinct situations of different groups and individuals whose human rights have been violated. Also, it would be much more inclusive than a state-centered approach as its focus is on the victims of human rights violations rather than on the agents of persecution.

The resulting expanded reading of a refugee as a person on the move who finds herself in a vulnerable situation may represent an essential epistemic resource for redesigning refugee governance and overcoming the centrality of states in the current geopolitical context. Not only might it pave the way to recognize a new legal subjectivity (i.e., climate refugees), but it might also dismantle the (almost) absolute role of the State in immigration policies.

This aspect leads us to the second step of this analysis: once decolonized the definition of a refugee, the unit of responsibility should be shifted from states to non-state actors.

The main reason for this shift is that states are not the most suitable subjects for doing environmental justice for "climate refugees." Even if they could successfully coordinate their joint action to cope with this challenge, they would not be interested in doing justice. Rather, consistent with the underlying triage governmentality, their primary aim would be to efficiently allocate scarce resources and sort efficient *converters* into economic advantage. By contrast, non-state actors acting together in a coordinated process would potentially accommodate different vulnerabilities at the individual and community level while implementing context-tailored policies.

This potential might be unlocked if their "responsibility" for doing environmental justice for "climate refugees" is conceived as "empowerment" rather than "liability." Indeed, the third step of this analysis implies a rethinking of the meaning of responsibility.

So far, the responsibility has been considered in terms of liability, thus requiring to demonstrate (1) a causality link between human mobility and climate change and (2) the link between the event (climate change) and agents. As mentioned earlier, both aspects remain, however, unresolved because of the weakness of causal attribution. The first aspect is still contested because migration is primarily understood as a multi-causal phenomenon as the environmental factor is not the "only driver in town." According to the 2011 Foresight Report, the drivers of migration are political, demographic, economic, social, and environmental (Foresight, 2011, p. 109). The second aspect is also problematic as it is often challenging to distinguish between anthropogenic climate change and nature-made environmental degradation. If the link between human activities and environmental degradation is not sufficiently demonstrated, one can label it as a natural disaster. While this latter is also likely to trigger human mobility, it does not imply any states' liability. In this regard, states may also object that they are not liable for anthropogenic climate change, as non-state actors like companies or multinationals are the "true" polluters instead. Ultimately, climate change is an example of a structural injustice produced by a plurality of subjects. Therefore, it is problematic to isolate and punish the perpetrators by a causal connection linking individual agents and the action at stake (Young, 2006).

A further limitation in conceiving responsibility as liability relies on its overly narrow focus on a backward-looking approach. This aspect neglects that EJ includes multiple *temporalities*, thus bridging past, present, and future.[20] Above all, if we understand the issue of "climate refugees" in the present-future-conditional tense, there is no sense to look at past injustices only. The right questions should be then: Who is taking care of "climate refugees"? Who is doing environmental justice for them? Which type of responsibility should we employ?

In answering those questions, I argue that non-state actors could better do environmental justice for "climate refugees" if they employ a forward-looking collective responsibility. This responsibility is *collective* because it is of *all* and will affect *all*, although differently. Major vulnerabilities suffered by "climate refugees" lacking a legal status depend upon the interplay of a plurality of subjects, thus reflecting a structural injustice. In this context, the mission of EJ is to dismantle the mechanisms by which the triage system governed by the capital and state disproportionately displaces the social and ecological costs to certain groups of people: "climate refugees." Finally, this responsibility is *forward-looking* because the primary aim is not to establish who is liable for actions committed in the past, still producing damages. Rather, the main goal is to identify a plurality of collective subjects having both the capacity and willingness to pursue a future environmental just-outcome. In the words of Neuhäuser, "the objective is to develop a normative approach that helps actors to collectively embrace their responsibility for future equitable conditions" (Neuhäuser, 2014, p. 233).

To this end, a different model of responsibility is needed in the first place: the Forward-Looking Collective Responsibility (FLCR). FLCR means "to make sense of a group *doing something* in the world and *taking responsibility* for bringing about a state of affairs" (Smiley, 2014, p. 3). In this case, I seek to direct FLCR at non-state actors (collective subjects), while the state of affairs to be pursued (collective aim) is environmental justice for "climate refugees": fostering their capabilities instead of treating them as commodities. In other words, collective subjects have to work together to allow "climate refugees" to rebuild a social world where they can flourish.

Secondly, a collective answer is needed. A collective answer does not mean that every state has to do its own part, as the principle Common but Differentiated Responsibilities and Respective Capabilities (CBDR – RC) would suggest in the frame of a state-centric approach.[21] Rather, a collective answer means the necessity to rethink the coordination of global collective action by assigning collective capabilities and specific functions to collective subjects.

But what is the meaning of collective subjects? Who are they? What are their collective capabilities and specific functions? For this analysis, I conceive a collective subject as a supra-individual subject: something more than a mere sum of individuals. It consists of "any group acting to make a right working in the concrete life of both the group as a collective subject and its members, improving their living conditions" (Rosignoli, 2018, p. 820). It may include any form of organized collectivities acting for making a right working in the concrete life of both the group and its members. Examples include non-governmental organizations (NGOs), multinational enterprises (MNEs), *sui generis* entities like the International Committee of the Red Cross and Red Crescent Movement (ICRC), and faith-based organizations.

In practical terms, climate-induced migration shall require international cooperation between collective subjects operating in countries of origin, transit, and destination. An FLCR model specifies what each collective subject should be doing to cope with "climate refugees" in the future global scenario. What they should be doing consists of their collective capabilities. Those are capabilities exercised by collective subjects acting "in order to secure a capability for the members of that group" (Robeyns, 2017, p. 116).

Collective capabilities are horizontal tools enabling an answer from below. Two types of collective capabilities have been categorized so far: resistant and resilient capabilities (Rosignoli, 2018).

I argue that resistant capabilities are the group's collective capacity to resist structural injustices (Rosignoli, 2018, p. 822). As for the case of climate-induced migration, resistant capabilities of non-state actors may include the following activities: building sea walls, dams, seawater desalination plants, artificial coral reefs, raising of the ground level, protests, social mobilization, making their voices heard in the international public spheres where they matter, and all activities aimed at preventing the territory being uninhabitable.

Resilient capabilities are defined as the "capacity of any group to react constructively to structural injustices" (Rosignoli, 2018, p. 822). Non-state actors' initiatives within this category may include adaptive strategies for those who decide to stay/ are unable to leave and joint actions for those who decide/are forced to migrate.

For those who decide to stay or are unable to leave, collective subjects exercise their resilient capabilities when organizing self-help initiatives and collective agency to rebuild a socio-environmentally habitable place. Such initiatives may require: building floating platforms and/or artificial islands; activating civil protection/Red Cross mechanisms to strengthen cooperation between countries of origin, transit, and destination, thus improving response to disasters; and coordination of third sector associations to support the most vulnerable groups.

Resilient capabilities of collective subjects for those who are forced to migrate may comprise activities to foster collaboration between schools, universities, trade unions, multinationals, NGOs, non-profit organizations, faith-based organizations, and so forth.

Examples of such activities are the following:

(1) facilitating migrants' children integration at schools;
(2) transferring people the necessary skills to adapt to market's labor of destination countries;
(3) identifying local knowledge/transferable skills which will significantly improve the placement potential of migrants;
(4) promoting programs of mutual learning and exchanges of good practice;
(5) informing migrants of rights they enjoy in destination countries, and providing contacts of legal advisers and lawyers offering unpaid legal assistance (e.g., NGOs, non-profit organizations providing free legal advice, local trade unions – if any – International Labor Organization);
(6) improving the cooperation between health authorities concerning pathologies and diseases (including mental diseases) existing in countries of origin and their medical treatments;
(7) involving faith-based organizations to facilitate accommodation finding and network with compatriots within countries of destination;
(8) encouraging multinationals to employ/train disadvantaged groups by relying on sustainability reporting tools (e.g., Global Reporting Initiative, GRI) and corporate social responsibility mechanisms; and
(9) involving professional and business organizations, if any, for shifting production of survival goods to cope with the emergency.

Concluding remarks

This chapter has analyzed the implications of "climate refugees" for (Decolonial) Environmental Justice. As outlined earlier, "climate refugees" are mostly citizens of developing countries. When displaced in the context of climate change, they are expected to experience three scenarios of injustice: (1) they do not cross any national border, thus becoming Internally Displaced Persons (IDPs), (2) they move to developing neighbor states, and (3) they end up in immigration detention camps. Whether directly impacted by climate change or indirectly by migration flows from neighbor states, the most vulnerable developing states are disproportionately affected by climate-induced migration. Likewise, people on the move are adversely impacted by survival goods' scarcity and subjected to violence throughout the migration process. The roots of this injustice rely on the fact that the so-called climate refugees are still not recognized at the international refugee law level.

In answering the question of what precisely is the "injustice" where people forced to migrate because of environmental disruptions are concerned, the chapter proceeds to explain why it is a matter of environmental justice (EJ) and what EJ is all about.

Drawing on a decolonial environmental justice perspective, it examines the threefold environmental injustice of "climate refugees" through the tripartite notion of coloniality to uncover, and try to overcome, the coloniality that prevents the recognition of climate refugees as new legal subjectivities.

The chapter has then shown that the operationalization of EJ can significantly vary depending on a state or a non-state-centric approach. If the unit of responsibility is the state, the EJ's original aim risks being misinterpreted, thus abdicating its radical mission to dismantle state and capital mechanisms producing structural injustices. In a state-centric perspective, "climate refugees" are conceived as undesirable/unwanted material flows producing environmental conflicts. Like waste whose generation is to be either prevented at source or equally distributed, "climate refugees" are more likely to be governed by principles of triage.

By contrast, a shift from the state to non-state actors as a unit of responsibility may give more chance to EJ to rethink the "climate refugees" governance in a more environmentally just light. From this different perspective, the term "responsibility" is not to be conceived as "accountability" for past emissions, but rather as "empowerment" for future challenges. With a multi-temporal scale including past, present, and future, a Forward-Looking Collective Responsibility model is then proposed as an alternative theoretical framework.

Following this model, the governance of climate-induced migration should be ruled by a plurality of collective subjects through their collective capabilities. Both resistant (i.e., activities aimed at preventing the territory from being uninhabitable) and resilient collective capabilities (i.e., collective agency to rebuild a socio-environmentally habitable place) may design an alternative future scenario for "climate refugees" that seems more likely to achieve the radical outcome EJ is expected to pursue.

Notes

1 Until 2020, the African Union Convention for the Protection and Assistance of Internally Displaced Persons in Africa (Kampala Convention) has been ratified by 31 of the African Union's 55 member states.
2 See Article 5(4) of the African Union, *African Union Convention for the Protection and Assistance of Internally Displaced Persons in Africa ("Kampala Convention")*, 23 October 2009, available at: www.refworld.org/docid/4ae572d82.html [accessed 20 October 2021].
3 For a detailed description of Australia's Naru, see Doherty, B. (2020, February 15). Australia's offshore detention is unlawful, says international criminal court prosecutor. *The Guardian*, available at www.theguardian.com/australia-news/2020/feb/15/australias-offshore-detention-is-unlawful-says-international-criminal-court-prosecutor [accessed 21 October 2021]; for a critical overview of violence perpetuated in immigration detention camps, see (Perocco, 2019).
4 Cf US Environmental Protection Agency's definition of environmental justice in the United States Environmental Protection Agency, 'Learn About Environmental Justice' www.epa.gov/environmentaljustice/learn-about-environmental-justice [accessed 25 October 2021]: "the fair treatment and meaningful involvement of all people regardless of race, color, national origin, or income with respect to the development, implementation, and enforcement of environmental laws, regulations, and policies."
5 In Latin America, the concept of environmental justice has been framed under the banner of Decolonial Environmental Justice (Rodriguez, 2020). In this regard, see also (Wald et al., 2019) underling how environmental justice as a framework cannot accommodate the different nuances of environmentalism introduced by indigenous activists and Decolonial Latin American Studies. According to this strand of research, even the term "justice" could not fully engage the ways that colonization persists by colonial values (coloniality), thus suggesting the expression "Latinx decolonial environmentalisms" as a new frontier of research.
6 For a deeper framework analysis of the "climate refugees" see the four framings discussed by Cooper et al. (victim, security threat, adaptive agent and political subject) (Ransan-Cooper, Farbotko, McNamara, Thornton, & Chevalier, 2015); cf. also (Fröhlich, 2017, p. 8): "The analysis of colonial stereotypes in European discourses on migration seems to be particularly fruitful with regard to media discourses in different European states, which depict 'the migrant', 'the refugee', 'the asylum seeker' in a certain way and thereby uncover colonial thought processes. The most striking ones are representations of migrants as 'victims' or as 'threats'; both perpetuate colonial power relations between seemingly stable, neutral, European identities and systems and threatening and/or damaged migrants standing outside of those spaces."
7 Material well-being is intended here as access to safe drinking water, clean air, wood, etc.
8 Extensively used by Giovanna di Chiro (Di Chiro, 2008), the term "social reproduction" refers to natural elements relevant to cultural heritage, spiritual practice, and the maintenance of everyday life.
9 To date, 26 European countries, mostly but not exclusively EU member states, have joined the Schengen Area. For a detailed list of Schengen countries, see www.schengenvisainfo.com/schengen-visa-countries-list/ [accessed 26 October 2021].
10 For a comprehensive overview of migration schemes related to environmental change, see also (IOM, 2019, p. 158-ss.).
11 For a critical examination of temporary work programs to New Zealand and Australia see (Klepp & Herbeck, 2016, pp. 67–69). Among others, the authors mention: Pacific Access Category (PAC), Recognized Seasonal Employment (RSE), Kiribati Australia Nursing Initiative (KANI). Unlike the EU and Australia, the form of temporary work offered in New Zealand often allows settling permanently later on.
12 Cf. (Hathaway & Neve, 1997, p. 120-ss).
13 (Adapted by Perchinig, 2017, p. 147).

14 The conditions of scarcity, triggered by economic stagnation and unemployment suffered by developed countries in the aftermath of the 2007–08 financial crisis and the 2020 Covid-19 pandemic, have been used to set priority rules. For a critical overview on how the pandemic has inverted previous hierarchies of more and less desired migrant workers, see (Triandafyllidou & Nalbandian, 2020).

15 As for labeling and social hierarchy, see also Gasper and Truong distinction: "Categorical boundaries are drawn to distinguish between: 'political refugees,' who may, at least in principle, be granted entry as exemplification of the political principles that the state endorses; 'economic refugees', whose claims for a right to asylum are denied, but who may if very fortunate be admitted under the next category; legal migrant workers, those invited in for furtherance of the national (economic) interest, on either temporary or indefinite terms, who are typically correspondingly subdivided into unskilled and skilled categories; and illegal migrant workers, who are at risk of deportation. Lastly, persons who have been trafficked under false pretences or coercion are officially protected by international law, but are often treated as illegal migrants and have their human rights ignored from all sides (GAATW, 2007)" (Gasper & Truong, 2010, p. 345).

16 I want to express my gratitude to lawyers like Alba Ferretti, Veronica Dini, Luca Saltalamacchia, and other members of the "Legalità per il clima" network for sharing their experiences and information while writing this chapter.

17 The interview was published by Capitalism Nature Socialism Journal in 2017.

18 Lee clarifies that the link between a conflict situation and environment issues can take on a variety of configurations depending on how the links are grouped. Given that some relate to causes and others to effects, these configurations result in clusters able to show the distribution of these links to continent place (Lee, 2020).

19 On the hot-spots potential to identify rights bearers, see also (Biermann & Boas, 2008; Kälin, 2010; Kent & Behrman, 2018).

20 Among others, see (David Pellow, 2017).

21 The Common but Differentiated Responsibilities and Respective Capabilities (CBDR – RC) principle has been laid down by the United Nations Framework Convention on Climate Change (UNFCCC). It aims to acknowledge the different capabilities and differing responsibilities of *individual countries* in addressing climate change.

Reference list

Agrawal, A., & Lemos, M. C. (2015). Adaptive development. *Nature Climate Change,* *5*(3), 185–187. https://doi.org/10.1038/nclimate2501

Agyeman, J., & Evans, B. (2004). "Just sustainability": The emerging discourse of environmental justice in Britain? *Geographical Journal, 170*(2), 155–164. https://doi.org/10.1111/j.0016-7398.2004.00117.x

Álvarez, L., & Coolsaet, B. (2020). Decolonizing environmental justice studies: A Latin American perspective. *Capitalism Nature Socialism, 31*(2), 50–69.

Baldwin, A. (2013). Racialisation and the figure of the climate-change migrant. *Environment and Planning A, 45*(6), 1474–1490. https://doi.org/10.1068/a45388

Barca, S. (2014). Telling the right story: Environmental violence and liberation narratives. *Environment and History, 20*(4), 535–546. https://doi.org/10.3197/096734014X140 91313617325

Basu, P. (2018). Environmental justice in South and Southeast Asia. In R. Holifield, J. Chakraborty, & G. Walker (Eds.), *The Routledge handbook of environmental justice* (pp. 603–614). https://doi.org/10.4324/9781315678986-48

Benford, R. D. (2005). The half-life of the environmental justice frame: Innovation, diffusion, and stagnation. In D. N. Pellow & R. J. Brulle (Eds.), *Power, justice and the environment* (pp. 37–54). London and Cambridge, MA: MIT Press.

Bettini, G. (2013). Climate barbarians at the gate? A critique of apocalyptic narratives on "climate refugees." *Geoforum, 45*, 63–72. https://doi.org/10.1016/j.geoforum.2012. 09.009

Bettini, G. (2014). Climate migration as an adaption strategy: De-securitizing climate-induced migration or making the unruly governable? *Critical Studies on Security, 2*(2), 180–195. https://doi.org/10.1080/21624887.2014.909225

Bettini, G., Nash, S. L., & Gioli, G. (2017). One step forward, two steps back? The fading contours of (in)justice in competing discourses on climate migration. *Geographical Journal, 183*(4), 348–358. https://doi.org/10.1111/geoj.12192

Biermann, F., & Boas, I. (2008). Protecting climate refugees: The case for a global protocol. *Environment, 50*(6), 8–17. https://doi.org/10.3200/ENVT.50.6.8-17

Bullard, R. (1990). *Dumping in dixie: Race, class, and environmental quality*. Boulder, CO: Westview Press.

Bullard, R. (1993). *Confronting environmental racism: Voices from the grassroots*. Boston: South End Press.

Bullard, R. (1994). *Unequal protection: Environmental justice and communities of color*. San Francisco, CA: Sierra Club Books.

Butler, J., & Athanasiou, A. (2015). *Dispossession the performative in the political: Conversations with Athena Athanasiou*. Cambridge: Polity.

Carruthers, D. V. (2008). *Environmental justice in Latin America: Problems, promise, and practice*. Cambridge, MA: MIT Press.

Castles, S. (2006a). *Back to the future? Can Europe meet its labour needs through migration? University of Oxford*. Oxford: International Migration Institute.

Castles, S. (2006b). Global perspectives on forced migration. *Asian and Pacific Migration Journal, 15*(1), 7–28. https://doi.org/10.1177/011719680601500102

Cutter, S. L. (2001). *American Hazardscapes: The regionalization hazards and disasters*. Washington, DC: Joseph Henry Press.

Di Chiro, G. (2008). Living environmentalisms: Coalition politics, social reproduction, and environmental justice. *Environmental Politics, 17*(2), 276–298. https://doi.org/10.1080/09644010801936230

Escobar, A. (2007). Worlds and knowledges otherwise: The Latin American modernity/ coloniality research program. *Cultural Studies, 21*(2–3), 179–210. https://doi.org/10.1080/09502380601162506

Faber, D. (1998). *The struggle for ecological democracy: Environmental justice movements in the United States, democracy and ecology* (D. Faber, Ed.). New York: Guilford Press.

Fan, M.-F., & Chou, K.-T. (2018). Environmental justice in a transitional and transboundary context in East Asia. In R. Holifield, J. Chakraborty, & G. Walker (Eds.), *The Routledge handbook of environmental justice* (pp. 615–626). Routledge. https://doi.org/10.4324/9781315678986-49

Fanon, F. (1967). *Black skin, White masks*. New York: Grove Press.

Felli, R. (2013). Managing climate insecurity by ensuring continuous capital accumulation: "Climate refugees" and "climate migrants." *New Political Economy, 18*(3), 337–363. https://doi.org/10.1080/13563467.2012.687716

Foresight: Migration and Global Environmental Change. (2011). *Final project report*. London: The government office for science. Retrieved from https://www.un.org/development/desa/pd/sites/www.un.org.development.desa.pd/files/unpd-cm10201202-11-1116-migration-and-global-environmental-change.pdf

Fröhlich, C. (2017). A critical view on human mobility in times of crisis. *Global Policy, 8*, 5–11. https://doi.org/10.1111/1758-5899.12417

Gasper, D., & Truong, T. D. (2010). Movements of the 'we': International and transnational migration and the capabilities approach. *Journal of Human Development and Capabilities, 11*(2), 339–357. https://doi.org/10.1080/19452821003677319

Giorgi, F. (2006). Climate change hot-spots. *Geophysical Research Letters, 33*(8), 1–4. https://doi.org/https://doi.org/10.1029/2006GL025734

Grosfoguel, R. (2007). The epistemic decolonial turn: Beyond political-economy paradigms. *Cultural Studies, 21*(2–3), 211–223. https://doi.org/10.1080/09502380601162514

Hathaway, J., & Foster, M. (2015). *The law of refugee status*. Cambridge: Cambridge University Press.

Hathaway, J. C., & Neve, R. A. (1997). Making international refugee law relevant again: A proposal for collectivized and solution-oriented protection. *Harvard Human Rights Journal, 10*, 115–212.

Holifield, R., Chakraborty, J., & Walker, G. (2017). *The Routledge handbook of environmental justice*. Routledge. https://doi.org/10.4324/9781315678986

Ide, T., Brzoska, M., Donges, J. F., & Schleussner, C. F. (2020). Multi-method evidence for when and how climate-related disasters contribute to armed conflict risk. *Global Environmental Change, 62*, 102063. https://doi.org/10.1016/j.gloenvcha.2020.102063

IDMC. (2021). *Global report on internal displacement*. Geneva: IDMC.

International Organization for Migration. (1996, April 21–24). *Environmentally-induced population displacements and environmental impacts resulting from mass migrations: International symposium*. Geneva: International Organization for Migration.

IOM. (2019). *World migration report 2020*. Geneva: IOM. Retrieved from https://publications.iom.int/system/files/pdf/wmr_2020.pdf

Juthe, A. (2016). Classifications of arguments by analogy part I. A comprehensive review of proposals for classifying arguments by analogy. *Cogency, 8*(2), 51–99.

Kahanec, M., & Zimmermann, K. F. (2012). High-skilled immigration policy in Europe. *SSRN Electronic Journal*. https://doi.org/10.2139/ssrn.1767902

Kälin, W. (2010). Conceptualising climate-induced displacement. In J. McAdam (Ed.), *Climate change and displacement: Multidisciplinary perspectives* (pp. 81–104). Oxford: Hart Publishing.

Kälin, W., & Schrepfer, N. (2012). *Protecting people crossing borders in the context of climate change: Normative gaps and possible approaches*. Geneva: UNHCR.

Kent, A., & Behrman, S. (2018). *Facilitating the resettlement and rights of climate refugees: An argument for developing existing principles and practices*. Abingdon: Routledge. https://doi.org/10.4324/9781351175708

Keucheyan, R. (2016). *Nature is a battlefield: Towards a political ecology*. Cambridge: Polity Press.

Klepp, S., & Herbeck, J. (2016). The politics of environmental migration and climate justice in the pacific region. *Journal of Human Rights and the Environment, 7*(1), 54–73. https://doi.org/10.4337/jhre.2016.01.03

Kukathas, C. (2016). Are refugees special? In S. Fine & L. Ypi (Eds.), *Migration in political theory* (pp. 249–268). https://doi.org/10.1093/acprof:oso/9780199676606.003.0012

Laurent, E. (2011). Issues in environmental justice within the European union. *Ecological Economics, 70*(11), 1846–1853. https://doi.org/10.1016/j.ecolecon.2011.06.025

Lee, J. R. (2009). *Climate change and armed conflict: Hot and cold wars*. London: Routledge.

Lee, J. R. (2020). *Environmental conflict and cooperation: Premise, purpose, persuasion, and promise*. Abingdon: Routledge.

Maldonado-Torres, N. (2007). On the coloniality of being: Contributions to the development of a concept. *Cultural Studies, 21*(2–3), 240–270. https://doi.org/10.1080/09502380601162548

Maldonado-Torres, N. (2011). Thinking through the decolonial turn: Post-continental interventions in theory, philosophy, and critique – an introduction. *Transmodernity: Journal of Peripheral Cultural Production of the Luso-Hispanic World, 1*(2), 1–15.

McGurty, E. M. (2000). Warren county, NC, and the emergence of the environmental justice movement: Unlikely coalitions and shared meanings in local collective action. *Society and Natural Resources, 13*(4), 373–387. https://doi.org/10.1080/089419200279027

McKinnon, C. (2012). *Climate change and future justice: Precaution, compensation and triage*. London: Routledge.

Mignolo, W. D. (2012). *Local histories/global designs: Coloniality, subaltern knowledges, and border thinking*. Princeton: Princeton University Press. https://doi.org/10.1215/0961754x-9-3-551

Moore, B. (2017). Refugee settlements and sustainable planning. *Forced Migration Review, 55*, 5–7.

Neuhäuser, C. (2014). Structural injustice and the distribution of forward-looking responsibility. *Midwest Studies in Philosophy, 38*(1), 232–251. https://doi.org/10.1111/misp.12026

Parenti, C. (2011). *Tropic of chaos: Climate change and the new geography of violence*. New York: Nation Books.

Park, L. S.-H., & Pellow, D. (2019). Forum 4: The environmental privilege of borders in the anthropocene. *Mobilities, 14*(3), 395–400. https://doi.org/10.1080/17450101.2019.1601397

Pellow, D. (2016). Toward a critical environmental justice studies: Black lives matter as an environmental justice challenge-corrigendum. *Du Bois Review, 13*(2), 1–16. https://doi.org/http://dx.doi.org/10.1017/S1742058X16000175

Pellow, D. (2017). *What is critical environmental justice?* Cambridge: Polity.

Pellow, D., & Brulle, R. (2005). Power, justice, and the environment: Toward critical environmental justice studies. *Power*. https://doi.org/10.1111/j.1600-0714.2010.00943.x

Perchinig, B. (2017). The challenge of migration for crisis and disaster management: Key concepts and recommendations. In F. Altenburg, A. Faustmann, T. Pfeffer, I. Skrivanek, & G. Biffl (Eds.), *Migration und Globalisierung in Zeiten des Umbruchs Festschrift für Gudrun Biffl* (pp. 135–153). Krems: Edition Donau-Universität Krems.

Perocco, F. (Ed.). (2019). *Tortura e migrazioni Torture and migration*. Venezia: Edizioni Ca' Foscari.

Phillips, A. S., Hung, Y.-T., & Bosela, P. A. (2007). Love canal tragedy. *Journal of Performance of Constructed Facilities, 21*(4), 313–319. https://doi.org/10.1061/(ASCE)0887-3828(2007)21:4(313)

Piguet, E. (2019). Theories of voluntary and forced migration. In *Routledge handbook of environmental displacement and migration* (pp. 17–28). https://doi.org/10.4324/9781315638843-2

Pulido, L. (1996). A critical review of the methodology of environmental racism research. *Antipode, 28*(2), 142–159.

Pulido, L. (2017). Conversations in environmental justice: An interview with David Pellow. *Capitalism, Nature, Socialism, 28*(2), 43–53. https://doi.org/10.1080/10455752.2016.1273963

Quijano, A. (2000). Coloniality of power and eurocentrism in Latin America. *International Sociology, 15*(2), 215–232.

Ransan-Cooper, H., Farbotko, C., McNamara, K. E., Thornton, F., & Chevalier, E. (2015). Being(s) framed: The means and ends of framing environmental migrants. *Global Environmental Change, 35*, 106–115. https://doi.org/10.1016/j.gloenvcha.2015.07.013

Robeyns, I. (2017). *Wellbeing, freedom and social justice – the capability approach reexamined*. Cambridge, UK: Open Book Publishers.

Rodriguez, I. (2020). Latin American decolonial environmental justice. In B. Coolsaet (Ed.), *Environmental justice key issues* (pp. 78–93). Milton: Routledge.

Rosignoli, F. (2018). Categorizing collective capabilities. *Partecipazione e Conflitto, 11*(3), 813–837. https://doi.org/10.1285/i20356609v11i3p813

Schlosberg, D. (2007). *Defining environmental justice: Theories, movements, and nature.* Oxford: Oxford Univ. Press.

Schlosberg, D., Rickards, L., & Byrne, J. (2018). Environmental justice and attachment to place. In R. Holifield, J. Chakraborty, & G. Walker (Eds.), *The Routledge handbook of environmental justice* (pp. 591–602). https://doi.org/10.4324/9781315678986-47

Sen, A. (1999). *Development as freedom.* Oxford: Oxford University Press. https://doi.org/10.1215/0961754X-9-2-350

Shaw, A. (2018). Environmental justice for a changing Arctic and its original peoples. In R. Holifield, J. Chakraborty, & G. Walker (Eds.), *The Routledge handbook of environmental justice* (pp. 504–514). https://doi.org/10.4324/9781315678986-40

Smiley, M. (2014). Future-looking collective responsibility: A preliminary analysis. *Midwest Studies in Philosophy, 38*(1), 1–11. https://doi.org/10.1111/misp.12012

Ssebunya, M., Morgan, S. N., & Okyere-Manu, B. D. (2019). Environmental justice: Towards an African perspective. In M. Chemhuru (Ed.), *African environmental ethics* (pp. 175–189). Springer. https://doi.org/10.1007/978-3-030-18807-8_12

Taylor, D. E. (2000). The rise of the environmental justice paradigm. *American Behavioral Scientist, 43*(4), 508–580. https://doi.org/10.1177/0002764200043004003

Triandafyllidou, A., & Nalbandian, L. (2020). *"Disposable" and "essential": Changes in the global hierarchies of migrant workers after COVID-19.* Geneva: IOM. Retrieved from https://publications.iom.int/system/files/pdf/disposable-and-essential.pdf

UNEP. (2007). *The global environment outlook 4 (GEO-4).* New York: UNEP.

UN High Commissioner for Refugees (UNHCR). (1998, 22 July). *Guiding Principles on Internal Displacement,* ADM 1.1, PRL 12.1, PR00/98/109. Retrieved January 8, 2022 from https://www.refworld.org/docid/3c3da07f7.html

United Church of Christ Commission for Racial Justice. (1987). *Toxic wastes and race in the United States: A national report on the racial and socio-economic characteristics of communities with Hazardous waste sites.* New York: United Church of Christ Commission for Racial Justice. Retrieved from https://www.nrc.gov/docs/ML1310/ML13109A339.pdf

United States General Accounting Office. (1983). *Siting of Hazardous waste landfills and their correlation with racial and economic status of surrounding communities.* Washington, DC: The Office.

Wald, S. D., Vázquez, D. J., Ybarra, P. S., Ray, S. J., Pulido, L., & Alaimo, S. (2019). *Latinx environmentalisms: Place, justice, and the decolonial.* Philadelphia: Temple University Press.

Westra, L. (2009). *Environmental justice and the rights of ecological refugees.* London: Earthscan. https://doi.org/10.4324/9781849770088

Whyte, K. (2017). Is it colonial déjà vu? Indigenous peoples and climate injustice. In *Humanities for the environment: Integrating knowledge, forging new constellations of practice.* London and New York: Routledge. https://doi.org/10.4324/9781315642659

Whyte, K., Talley, J., & Gibson, J. (2019). Indigenous mobility traditions, colonialism, and the anthropocene. *Mobilities, 14*(3), 319–335. https://doi.org/10.1080/17450101.2019.1611015

Young, I. M. (2006). Responsibility and global justice: A social connection model. *Social Philosophy and Policy, 23*(1), 102. https://doi.org/10.1017/S0265052506060043

5 Environmental justice for "Climate Refugees"

Actors, instruments, and strategies

Why non-state actors can "solve" the justice dilemma

As argued in Chapter 4, the justice dilemma that surrounds the search for recognition of climate refugees may be better "solved" by non-state actors. The term "non-state actors" here includes non-governmental organizations (NGOs), civic associations, entities like the International Committee of the Red Cross and Red Crescent Movement (ICRC), and faith-based organizations. With this chapter, the analysis shifts from *why not* having states to *why having* non-state actors as key actors to do environmental justice for climate refugees.

To begin with, the higher suitability of these actors to pursue environmental justice is the result of various factors.

The first is that non-state actors may well enable the transition from a transnational to a translocal perspective. Following the "new" Hot Wars geopolitical context, climate-induced migration can no longer be seen through the "old" divide between developing and developed States, as this phenomenon is more likely to occur in the so-called climate hot-spots: regions that do not necessarily coincide with the borders of a single state. In the face of such climate hubs of migration and displacement, a translocal perspective facilitates actions to take into processes that trespass boundaries on different scales (Greiner & Sakdapolrak, 2013). Unlike the transnational approach still embedded in the idea of hierarchical and distinguishable scales, translocality does not mean adding another scale between the global and the local. Rather, it means overcoming this hierarchical logic by understanding the dynamic of various socio-spatialities beyond the national entities, nationalist historiographies, and Eurocentric view of global history (Greiner & Sakdapolrak, 2013; Verne, 2012). This translocal approach results then in something in-between "the local" and "the global" able to capture "the diverse and contradictory effects of interconnectedness between places, institutions and actors" (Greiner & Sakdapolrak, 2013, p. 375). This approach is particularly appropriate for dealing with environmental injustices linked with the lack of recognition of climate refugees. In this regard, the indigenous peoples' struggles in the context of the *Hot and Cold Wars* (Lee, 2009, 2020) represent a useful example for this analysis (Rodriguez, 2020; Schlosberg, Rickards, & Byrne, 2018; Shaw, 2018). On the one hand, indigenous peoples from the Global

DOI: 10.4324/9781003102632-5

South are fighting for new culturally differentiated forms of decision-making in nation-state models to acknowledge and preserve their cultural survival in the face of climate migration and displacement (Rodriguez, 2020). On the other hand, those in the Global North are becoming ever more translocal allies in resisting neighboring Arctic states competing for lands of new opportunities (Keucheyan, 2016; Whyte, 2017). A telling example of this translocal alliance is the Sámi: the indigenous peoples of Northern Europe that live in an area encompassing parts of Norway, Sweden, Finland, and the Kola Peninsula of Russia. It follows that Western categories like the nation-state do not fully represent the current reality of global civil society. By contrast, non-state actors may best challenge boundaries and hierarchy in scales, thus achieving the right balance between EJ claims from below and technical solutions from above across different localities regardless of its administration or national boundaries.

The second is that non-state actors' capacity to influence the global policy is getting ever more reliant on their ability to take climate actions also in the lack of a national mechanism in place (Pinto-Bazurco, 2018).[1] The intensified interplay between states and non-state actors and the role of orchestrator assigned to the UNFCCC within the *hybrid* architecture[2] of the Paris Agreement has further blurred the traditional categorizations of top-down and bottom-up (Kuyper, Linnér, & Schroeder, 2018), thus opening up new, translocal horizons to non-state actors. In light of this renovated global climate governance landscape, non-state actors are therefore in a better position for doing environmental justice for climate refugees due to the specific functions entrusted by the Paris Agreement. In particular, non-state actors are invited to act both as watchdogs of the nationally determined contributions (NDCs) and as governing partners by scaling-up their climate actions to be collected and shared in the Non-State Actor Zone for Climate Action platform (NAZCA) or through the Lima – Paris Action Agenda (LPAA) (Kuyper et al., 2018).[3] As a result, not only is the influence of non-state actors pivotal for their capacity to monitor, participate, and contribute to national climate actions, but is becoming increasingly relevant also in the governance beyond the state as it unfolds in climate actions pursued on their own, cross-border networks with peers around the world, and partnership with international organizations. To sum up, these specific functions and capabilities make non-state actors more equipped to deal with translocal climate migration and displacement both for resisting state inaction and for fostering adaptive strategies.

The third factor has to do with non-state actors' contribution in strengthening the democratic legitimacy of environmental governance through a better implementation of the principle of public participation. In line with the environmental justice-oriented aim of ensuring meaningful involvement of all in the decision-making process, non-state actors have the potential to fill the democratic deficit at the global policy level. In particular, non-state actors are more likely to give a voice to marginalized communities most affected by climate inaction, whose claims are more often ignored, silenced, or disregarded by policy-makers throughout the decision-making process concerning climate-induced migration and displacement.

A telling example is the local NGO based in Bougainville called Tulele Peisa that has organized and facilitated the voluntary relocation of indigenous communities (see Chapter 3). Tulele Peisa (Sailing the waves on our own) is considered one of the first organizations engaging community-driven climate refugee relocation efforts in the Pacific region (United Nations Development Programme, 2016). Indeed, not only non-state actors give a voice to unheard communities, but they can also build alternative policies and climate actions to resist or adapt climate disruptions regardless of the domestic politics. To this end, they usually form translocal alliances with peers worldwide while inserting themselves in international organizations' policy process. In doing so, they are more likely to influence the global agenda-setting, policy formulation, and decision-making within the UNFCCC orchestration.

To sum up, the role of non-state actors designed by the Paris Agreement has the potential to open up new spaces of public participation by bridging the micro- with meso-level and connecting the global and local levels.

A toolkit for non-state actors: collective capabilities

To date, the Paris Agreement has provided the necessary global framework for including non-state actors in the global climate governance orchestration. However, its hybrid architecture that ensures non-state actors' participation seems a necessary but not sufficient condition to make their actions consistent and truly effective to facilitate the construction of climate refugees as a new legal subjectivity.

To achieve this goal, non-state actors should be further equipped with collective capabilities.[4]

As described in Chapter 4, collective capabilities are defined as those capabilities exercised by a group that acts "to secure a capability for the members of that group" (Robeyns, 2017, p. 116). Categorized as resistant and resilient capabilities (Rosignoli, 2018), they are crucial for identifying and further developing strategies that non-state actors can adopt to make their collective actions effective in the spheres where they matter.

Drawing on (Williamson et al., 2020), I have selected and read the following Community Change Strategies in conjunction with the collective capabilities categorization and the functions non-state actors play within the Paris Agreement.

As shown in table 5.1, strengthening resistant capabilities means promoting strategies such as civil disobedience, climate litigation, and mass mobilizations. As pointed out by Williamson et al., civil disobedience is "the refusal to comply with certain laws or to pay taxes and fines, as a peaceful form of political protest, that often includes nonviolent techniques such as boycotting, picketing" (Williamson et al., 2020, p. 8). By scaling up this strategy, the resulting translocal disobedience is more likely to affect the global climate governance as non-state actors are crucial for implementation, and implementation is and will be key to do justice for climate refugees. In the light of translocality, non-state actors can be extremely relevant if their disobedience is "around the clock and around the globe," i.e., when coordination efforts lead to a domino effect in as many countries

Table 5.1 Collective capabilities, strategies, and functions of non-state actors.

Collective capabilities	Strategies	Functions
Resistant	Civil disobedience climate change litigation mass mobilization	Watchdogs (monitoring)
Resilient	Letter writing media advocacy policy advocacy partnership coalition-building	Governing partners (scale-up climate actions)

Source: Adapted from (Williamson et al., 2020).
www.mdpi.com/1660-4601/17/11/3765

as possible. In doing so, their coordinated actions may be useful for resisting effectively structural injustices or policies increasing environmental inequalities in the aftermath of large-scale environmental displacement (Jacobs, 2016).

Climate change litigation is a different strategy that non-state actors can put in place when states fail to comply with emissions reduction or limitation targets exacerbating climate-induced migration and displacement. This strategy fits well the translocal goal of creating a domino effect in several countries, thus increasing the pressure on governments and media advocacy (Konkes, 2018). The success of the Urgenda case perfectly embodies this approach, with numerous communities around the world having initiated proceedings against states to achieve similar rulings. As illustrated by the 2020 policy report on Global trends in climate change, there are ongoing legal proceedings in Ireland, France, Belgium, Sweden, Switzerland, Germany, the United States, Canada, Peru, and South Korea (Setzer & Byrnes, 2020). As observed by this report, climate change litigation would not only have the potential positive impact to bring non-compliant states in line with the Paris Agreement's temperature goal, but it might have an indirect, positive impact even when cases are unsuccessful. Indeed, the courts might dismiss plaintiffs' claims while nonetheless acknowledging the risks imposed by climate change and pave the way for future successes under different circumstances (Setzer & Byrnes, 2020). In this regard, a useful example is the Teitiota case. Although the Supreme Court of New Zealand dismissed his appeal against the decision to refuse him the refugee status, the court noted that environmental degradation resulting from climate change or other natural disasters could create a pathway into the Refugee Convention or protected person jurisdiction in an appropriate case. For example, if the applicant would face serious harm if returned, and there was evidence that his or her government was failing to take steps to protect its citizens from the effects of environmental degradation[5] (see also Chapter 2).

Finally, non-state actors may strengthen their resistant capabilities by organizing translocal, mass mobilizations. To name but one, the Global Climate March was one of the most successful mobilizations of this kind. Organized by a huge coalition of climate groups and activists on the eve of COP 21 in Paris, the Global

Climate March took place in various cities all over the world. On 28–29 November 2015, an estimated number of 785,000 people in 175 countries have been involved in 2,300 marches worldwide, thus leading to wide media coverage.[6] As mentioned earlier, the great convergence of all these groups on climate justice demands materialized, *inter alia*, in a direct reference to climate justice in the Paris Agreement's preamble.

These strategies may render non-state actors' monitoring function more effective in combating states' noncompliance within and outside the hybrid architecture of the Paris Agreement.

Climate change litigation, for example, may be a useful tool to preserve and enhance the contestatory practices of civil society.[7] In other words, resistant capabilities are crucial for avoiding that the orchestration UNFCCC model transforms non-state actors into willing servants, thus undermining their contestatory potential (Dryzek, 2017; Kuyper et al., 2018).

If non-state actors acted only within the hybrid architecture of the Paris Agreement, they would run the risk of reinforcing an unjust global order rather than challenging the status quo of distributive politics. Thus, expanding their resistant capabilities is necessary to prevent them from reproducing the same unfair state-centric outcomes at the global level.

As for resilient capabilities, there are several strategies that non-state actors may want to use to strengthen their ability to cope with the adverse impacts of structural injustices. As illustrated in table 5.1, those include letter writing, media advocacy, policy advocacy, partnership, and coalition building. Williamson et al., have described letter writing as "an organized effort to coordinate as many people as possible to write to a decision maker (legislative or facility) asking them to take a particular action" (Williamson et al., 2020, p. 8). Given their function as governing partners, non-state actors may well use this strategy to foster the scaling up of specific climate actions as an alternative to unfair policies eventually supported by the state. Although not all letter writing campaigns are successful, this strategy has a long history in the tradition of social movements and continues to be pursued online by email campaigns (Earl & Kimport, 2008).

Media advocacy may also be used to disseminate information and promote policy initiatives crucial for ensuring the successful implementation of strategic adaptation initiatives within societies.

In this regard, non-state actors may well pursue this strategy through traditional and social media platforms. Along with traditional media and communication strategies, such as press releases (international, national, and community based), public communication, posters, flyers, and placards, media advocacy has been increasingly advanced through cheaper and far-reaching social and visual media tools (Oyedemi, 2020). By their websites, blogs, visual media tools (photovoice), performances (community drama, theatre, flash mobs), and social media channels, non-state actors have significantly increased their capacity to make available information on policy initiatives to an ever-widening audience base. In doing so, not only is increased their domestic consent building effort, but also the translocal community support for their initiatives, including increased possibilities for scaling up best practices. This latter aspect well fits the idea of reacting constructively to structural injustices by linking media with policy advocacy. This implies to make as visible as possible the development of policy-based solutions to affect social change and provide for alternatives. To overcome the binarism between developed and developing states that may render non-state actions' collective action ineffective as they are too local or regional in their conceptual scope, establishing partnership and building coalition may facilitate the sharing of best practice, translocal coordination, and mutual technical assistance.

Limitations and ways forward

Non-state actors' mission to do environmental justice for climate refugees by strengthening their collective capabilities may be undermined by the lack of adequate financial and economic resources, knowledge, and technical expertise. These factors have the potential to limit their capacities to implement with success the strategies mentioned earlier.

In this context, a challenging issue remains how to overcome the divide between wealthier organizations in the Global North and those in the Global South. Such divide implies that the latter are usually less endowed in terms of staff, members, degree of professionalization, budget, and often face greater constraints on participation than their counterparts in the industrial democracies of the Global North (Carruthers, 2008). Although the capacity to influence the global policy by collective actions is not directly proportional to the availability of resources and memberships (Kousis, della Porta, & Jiménez-Sánchez, 2008), the gap in environmental organization's resources may polarize non-state actors along identities line. In turn, the manifold identities of climate groups may prevent them from creating translocal, effective alliances, thus weakening their joint actions and coordination efforts.

Further limitations have to do with threats to organizational independence. Challenges arise both when environmental organizations can overly benefit from state funding and when they cannot or only to a limited extent.

In the first case, environmental organizations risk being too reliant on the state's funding. This may inhibit their capacity to resist and even criticize unfair or unsustainable policies supported by the state. In other words, all strategies discussed earlier concerning resistant capabilities might be affected when environmental organizations are not independent enough.

In the second case, environmental organizations that can receive a limited amount of or no resources from the state face other obstacles. To survive, they run the risk of becoming overly dependent on wealthy donors and or special-interest groups wanting services – that eventually are not core to their function – in exchange (Egdell & Dutton, 2017). Alternatively, they have to compete for access to other external funding. Even though the Paris Agreement has made the access to funds more transparent and reliable, it is often characterized by complex procedures, mechanisms, and required conditions that are likely to disadvantage certain countries at the expense of others.

In theory, the Paris Agreement has reframed the flow of finances according to the principle of Common but Differentiated Responsibilities (CBDR) by encouraging the transfer of resources and adequate financial resources from developed to developing states. A useful initiative to support developing states in understanding financial mechanisms is the Capacity Building Initiative for Transparency (CBIT) established by the Paris Agreement for building "institutional and technical" capacity in climate governance. In practice, however, concrete challenges stem from the lack of a definition of "climate finance" in the first place. As noted by Deane and Hamman, this absence is open to discretionary accounts "of how much is pledged (versus how much is actually 'financed'), through which sources (public, private, intermediaries, etc.) and what the purpose of the funds actually is" (Deane & Hamman, 2018, p. 76). Once countries have obtained funding, a further issue is monitoring that they manage and use the funds effectively. In this context, a key challenge is strengthening the capacity of developing countries to manage the funds properly, being the most affected by the impacts of climate change.

To conclude, whether resources stem from donors, private companies, or external funding, the disproportionate reliance on one of these funding arrangements and the limited capacity to manage funds effectively are likely to undermine their policy agenda and social mission.

The common denominator of both cases is that limited independence may lead to a process of mission drift (Egdell & Dutton, 2017). This implies that environmental organizations risk getting ever more responsive to their donors than civil society, thus losing enhanced trust and legitimacy (Jordan & van Tuijl, 2012). If so, the side effect would be for them to cultivate illegitimacy and mistrust, i.e., two sources of noncompliance and noncooperation that have been documented by the literature (Rodríguez-Muñiz, 2017). As a result, their role as governing partners would not be exercised effectively as local communities may perceive them as the independent voice of service users (Egdell & Dutton, 2017). At the same time, international organizations may lose interest in involving them in policy processes if not useful to spur implementation. Ultimately, a further unwanted effect may be the reinforcement rather than dismantlement of unequal power relations underpinning a global elite rather than a global civil society (Dryzek, 2012; Heins, 2005). This reinforcement may re-establish a top-down, hierarchical logic that betrays the mission of grassroots environmental organizations while failing to reflect the hybrid architecture of the Paris Agreement.

To overcome these limitations, translocal alliances and mutual technical assistance may help most disadvantaged organizations in strengthening their collective capabilities. Technical assistance (TA) is defined as tailored support that organizations receive from technical experts or more experienced peers to complete various tasks (Freudenberg, Pastor, & Israel, 2011). In this view, even when lacking financial or knowledge resources, environmental organizations can benefit from being part of strong networks with wealthier and high-skilled counterparts in other parts of the world. Thus, promoting translocal solidarity (Routledge, 2012) through mutual technical assistance may foster their willingness to receive TA. Indeed, some evidence shows that not all organizations are either ready or willing to receive TA (Freudenberg et al., 2011). Some of them are reluctant to receive TA if they suspect that revealing areas of weakness might diminish future funding opportunities (Mitchell, Florin, & Stevenson,

2002). To overcome this impasse, it may be fruitful for a TA either delivered by experienced peers as they face similar challenges in strengthening their collective capabilities or academic partners. Evidence shows that university facilitation using an action learning approach effectively increases knowledge to action (Bonney, Welter, Jarpe-Ratner, & Conroy, 2019). This kind of activity may also help develop familiarity with each partner, build coalitions, and foster more collaborative efforts (Mitchell et al., 2002).

Another crucial aspect of coping with the independence issue is achieving a meaningful diversification of funding to become less reliant on the state or any other supranational institutions. Crucially, non-state actors should learn to act as critical allies of the state without compromising their advocacy role. This means that they have to cooperate as trusted partners (only) when state-led actions meet justice demands. If those demands were ignored or unfair policies were pursued, non-state actors would have two strategies.

The first is disengaging from non-state consent-building work by creating counter-hegemonic narratives to shift the consensus toward alternative, sustainable solutions (Hagai, 2006). Given that consensus is essential to accompany the social change in any given policy area, they may contribute to building or unravel consensus depending on the fairness of state-led actions.

The second is to increase their efforts in influencing the different policy cycles of international organizations, such as agenda-setting, policy formulation, decision-making, and implementation of global policies. In other words, non-state actors may change allies by taking advantage of the role assigned by the Paris Agreement to spur implementation of its general goals.

While the first strategy concerns the micro-level and aims to build a counter-hegemonic resistance from below, the second has to do with the macro-level and takes advantage of hierarchical scales between the global and domestic levels, and non-state actors' crucial role in ensuring and facilitating implementation.

When non-state actors meet the independence requirement, they become the voice of the most vulnerable groups in the global policy arena, thus galvanizing their role in influencing the climate global policy toward a more community-sensitive approach. The more non-state actors are successful in pushing global policies to take into account communities' needs, the more effective the principle of public participation will be by extension.

Conclusion

What precisely is the "injustice" where people forced to migrate because of environmental disruptions are concerned? Who should be doing environmental justice for climate refugees? Which instruments and strategies better equip the actors selected for this mission? This book has tried to answer those questions by re-situating the debate on the search for recognition of climate refugees through the lens of environmental justice.

Having accomplished that, the shortcomings in international law ultimately stem from the coloniality of power, knowledge, and being, it has tried to decolonize the figure of climate refugees while paving a way out of the 'legal impasse.'

To this end, it has provided a broader interpretation of a refugee based on analogy by law anchored in the concept of vulnerability. Extensively interpreted as a person on the move who finds herself in a vulnerable situation, the figure of climate refugee has been used as an essential epistemic resource for redesigning refugee governance and overcoming the centrality of states in the current geopolitical context.

The broader interpretation of a refugee has been then complemented with the construction of a new subjectivity through the practices of non-state actors.

Unlike states, non-state actors best suit the "new' Hot Wars geopolitical context, where the "old" divide between developing and developed states has been overcome by the presence of numerous and diffused climate "hot-spots." Their translocal approach may further overcome the hierarchical logic employed by states, through a dynamic understanding of climate-induced migration beyond national entities, nationalist historiographies, and Eurocentric view of global history (Greiner & Sakdapolrak, 2013; Verne, 2012). In particular, specific functions and capabilities assigned within the framework of the Paris Agreement make non-state actors more equipped to deal with translocal climate migration and displacement both for resisting state inaction and for fostering adaptive strategies.

Although their roles within the Paris Agreement have the potential to open up new spaces of public participation by bridging the micro- with meso-level and connecting the global and local levels, an outstanding question in this analysis is: How to provide a collective answer for climate refugees' governance in a globalized world of asymmetrical capacities?

A final section has been dedicated to answering this question by critically reflecting on additional instruments and strategies that may help non-state actors to cope with this challenge when dealing with climate refugees. While these strategies are not comprehensive since future research on this topic is expected and strongly encouraged, the ultimate argument of this book is the following.

The necessary collective answer to govern climate-induced migration implies rethinking the coordination of global collective action by fostering collective capabilities and specific functions of non-state actors. Their role is and will be crucial to accommodate different forms of vulnerability at the individual and community level while implementing context-tailored policies to do environmental justice for climate refugees while facilitating the construction of a new legal subjectivity.

Notes

1 The centrality of cities in exceeding climate and energy targets was not unknown before the Paris Agreement adoption. Since the early 1990s, transnational city networks, such as the International Council for Local Environmental Initiatives (ICLEI) and the Climate Alliance, have been established for this purpose. The Covenant of Mayors, launched in 2008 in Europe, is one of the most ambitious initiatives to gather local government to achieve climate targets. It gathers more than 9,000 local and regional authorities across 57 countries (Gordon & Johnson, 2017). However, the withdrawal of the US from the Paris Agreement under the Trump administration has made more visible the essential role of local governments, industry leaders, and NGOs in developing strategies for technical innovation and fostering the transformation trend of the global climate governance regime regardless of domestic affairs (Zhang, Chao, Zheng, & Huang, 2017). Indeed, Trump's formal declaration of withdrawal has not prevented more than 200 city mayors,

companies, and NGOs from pursuing the Paris Agreement's goals within orchestration efforts by the UNFCCC.

2 The term "hybrid multilateralism" has been advanced to describe the interplay between states and non-state actors in the new landscape of international climate cooperation. This interplay has increasingly challenged the traditional categorizations of "top-down" and "bottom-up" initiatives, especially because of the expanded roles assigned to non-state actors by the Paris Agreement (Backstrand, Kuyper, Linnér, & Lövbrand, 2017).

3 On the roles of non-state actors in the Paris Agreement framework, see also (Van Asselt, 2016).

4 Initially elaborated by Amartya Sen's work (Sen, 1980, 1999) with further conceptualization by Martha Nussbaum (Nussbaum, 2000) and others (Holland, 2014; Robeyns, 2005), the Capability Approach (CA) has been used by EJ scholars as a theoretical framework to look at the environmental injustices. Examples range from struggles around nature conservation (Martin, 2017), energy justice (Day, Walker, & Simcock, 2016; Partridge, Thomas, Pidgeon, & Harthorn, 2018), climate change and adaptation (Holland, 2017; Schlosberg, 2012), and empirical applications exploring implications for the wellbeing of communities living close to industrially contaminated sites (Pasetto et al., 2020). For this analysis, see also the concept of community capabilities developed by Schlosberg and Carruthers. Defined as the "ability to continue and reproduce the traditions, practices, cosmologies, and the relationships with nature that tie native peoples to their ancestral lands" (Schlosberg & Carruthers, 2010, p. 13), it is linked with the idea of social reproduction in the context of indigenous struggles.

5 See *Ioane Teitiota v. The Chief Executive of the Ministry of Business, Innovation and Employment*, [2015] NZSC 107, New Zealand: Supreme Court, 20 July 2015, available at: www.refworld.org/cases,NZL_SC,55c8675d4.html [accessed 7 December 2020]. Cf. The UN Human Rights committee's decision of 7 January, 2020, in the case of Teitiota v. New Zealand. *Ioane Teitiota v. New Zealand (advance unedited version)*, CCPR/C/127/D/2728/2016, UN Human Rights Committee (HRC), 7 January 2020, available at: www.refworld.org/cases, HRC,5e26f7134.html [accessed 7 December 2020].

6 Cf. https://350.org/global-climate-march/.

7 For a critical examination of this topic, see (Kent & Behrman, 2020).

Reference list

Backstrand, K., Kuyper, J. W., Linnér, B.-O., & Lövbrand, E. (2017). Non-state actors in global climate governance: From Copenhagen to Paris and beyond. *Environmental Politics, 26*(4), 561–579. https://doi.org/10.1080/09644016.2017.1327485

Bonney, T., Welter, C., Jarpe-Ratner, E., & Conroy, L. M. (2019). Understanding the role of academic partners as technical assistance providers: Results from an exploratory study to address precarious work. *International Journal of Environmental Research and Public Health, 16*(20). https://doi.org/10.3390/ijerph16203903

Carruthers, D. V. (2008). *Environmental justice in Latin America: Problems, promise, and practice*. Cambridge, MA: MIT Press.

Day, R., Walker, G., & Simcock, N. (2016). Conceptualising energy use and energy poverty using a capabilities framework. *Energy Policy, 93*, 255–264. https://doi.org/10.1016/j.enpol.2016.03.019

Deane, F., & Hamman, E. (2018). Transparency in climate finance after Paris: Towards a more effective climate governance framework. In M. Rimmer (Ed.), *Intellectual property and clean energy: The Paris agreement and climate justice* (pp. 69–92). Singapore: Springer. https://doi.org/10.1007/978-981-13-2155-9_3

Dryzek, J. S. (2012). Global civil society: The progress of post-Westphalian politics. *Annual Review of Political Science.* https://doi.org/10.1146/annurev-polisci-042010-164946

Dryzek, J. S. (2017). The meanings of life for non-state actors in climate politics. *Environmental Politics, 26*(4), 789–799. https://doi.org/10.1080/09644016.2017.1321724

Earl, J., & Kimport, K. (2008). The targets of online protest: State and private targets of four online protest tactics. *Information Communication and Society, 11*(4), 449–472. https://doi.org/10.1080/13691180801999035

Egdell, V., & Dutton, M. (2017). Third sector independence: Relations with the state in an age of austerity. *Voluntary Sector Review, 8*(1), 25–40. https://doi.org/http://dx.doi.org/10.1332/204080516X14739278719772

Freudenberg, N., Pastor, M., & Israel, B. (2011). Strengthening community capacity to participate in making decisions to reduce disproportionate environmental exposures. *American Journal of Public Health, 101*(Suppl. 1), S123–S130. https://doi.org/10.2105/AJPH.2011.300265

Gordon, D. J., & Johnson, C. A. (2017). The orchestration of global urban climate governance: Conducting power in the post-Paris climate regime. *Environmental Politics, 26*(4), 694–714.

Greiner, C., & Sakdapolrak, P. (2013). Translocality: Concepts, applications and emerging research perspectives. *Geography Compass, 7*(5), 373–384. https://doi.org/10.1111/gec3.12048

Hagai, K. (2006). Gramsci, hegemony, and global civil society networks. *Voluntas: International Journal of Voluntary and Nonprofit Organizations, 17*(4), 333–348.

Heins, V. (2005). Global civil society as politics of faith. In G. Baker & D. Chandler (Eds.), *Global civil society: Contested futures* (pp. 186–201). London: Routledge. https://doi.org/10.4324/9780203001486

Holland, B. (2014). *Allocating the earth: A distributional framework for protecting capabilities in environmental law and policy* (1st ed.). Oxford: Oxford University Press.

Holland, B. (2017). Procedural justice in local climate adaptation. *Environmental Politics, 26*(3), 391–412.

Jacobs, M. (2016). High pressure for low emissions: How civil society created the Paris climate agreement. *Juncture, 22*(4), 314–323.

Jordan, L., & van Tuijl, P. (2012). *NGO accountability: Politics, principles and innovations* (L. Jordan & P. van Tuijl, Eds.). London: Earthscan. https://doi.org/10.4324/9781849772099

Kent, A., & Behrman, S. (2020). Climate-induced migration: Will tribunals save the day? *Hong Kong Journal of Law and Public Affairs, 2*(2). https://doi.org/http://dx.doi.org/10.2139/ssrn.3682504

Keucheyan, R. (2016). *Nature is a battlefield: Towards a political ecology.* Cambridge: Polity Press.

Konkes, C. (2018). Green lawfare: Environmental public interest litigation and mediatized environmental conflict. *Environmental Communication, 12*(2), 191–203.

Kousis, M., della porta, D., & Jiménez-Sánchez, M. (2008). *Southern European environmental movements in comparative perspective. American behavioral scientist – Amer Behav Sci* (Vol. 51). https://doi.org/10.1177/0002764208316361

Kuyper, J. W., Linnér, B. O., & Schroeder, H. (2018). Non-state actors in hybrid global climate governance: Justice, legitimacy, and effectiveness in a post-Paris era. *Wiley Interdisciplinary Reviews: Climate Change.* https://doi.org/10.1002/wcc.497

Lee, J. R. (2009). *Climate change and armed conflict: Hot and cold wars.* London: Routledge.

Lee, J. R. (2020). *Environmental conflict and cooperation: Premise, purpose, persuasion, and promise.* Abingdon: Routledge.

Martin, A. (2017). *Just conservation: Biodiversity, wellbeing and sustainability*. Abingdon, UK: Routledge. https://doi.org/10.4324/9781315765341

Mitchell, R. E., Florin, P., & Stevenson, J. F. (2002). Supporting community-based prevention and health promotion initiatives: Developing effective technical assistance systems. *Health Education & Behavior, 29*(5), 620–639. https://doi.org/10.1177/109019802237029

Nussbaum, M. (2000). *Women and human development: The capabilities approach. The John Robert Seeley lectures* (1. publ.). Cambridge: Cambridge University Press.

Oyedemi, T. D. (2020). Protest as communication for development and social change BT – handbook of communication for development and social change. In J. Servaes (Ed.) (pp. 615–632). Singapore: Springer Singapore. https://doi.org/10.1007/978-981-15-2014-3_132

Partridge, T., Thomas, M., Pidgeon, N., & Harthorn, B. H. (2018). Urgency in energy justice: Contestation and time in prospective shale extraction in the United States and United Kingdom. *Energy Research and Social Science, 42*, 138–146. https://doi.org/10.1016/j.erss.2018.03.018

Pasetto, R., Marsili, D., Rosignoli, F., Bisceglia, L., Caranci, N., Fabri, A., . . . Mannarini, T. (2020). Promozione della giustizia ambientale nei siti industriali contaminati. *Epidemiologia e Prevenzione, 44*(5–6), 417–425. https://doi.org/10.19191/EP20.5-6.A001

Pinto-Bazurco, J. F. (2018). Remarks of Jose Felix Pinto-Bazurco. *Proceedings of the ASIL Annual Meeting, 112*, 235–236. https://doi.org/DOI: 10.1017/amp.2018.18

Robeyns, I. (2005). The capability approach: A theoretical survey. *Journal of Human Development, 6*(1), 93–117. https://doi.org/10.1080/146498805200034266

Robeyns, I. (2017). *Wellbeing, freedom and social justice – the capability approach re-examined*. Cambridge, UK: Open Book Publishers.

Rodríguez-Muñiz, M. (2017). Cultivating consent: Nonstate leaders and the orchestration of state legibility. *American Journal of Sociology, 123*(2), 385–425.

Rodriguez, I. (2020). Latin American decolonial environmental justice. In B. Coolsaet (Ed.), *Environmental justice key issues* (pp. 78–93). Milton: Routledge.

Rosignoli, F. (2018). Categorizing collective capabilities. *Partecipazione e Conflitto, 11*(3), 813–837. https://doi.org/10.1285/i20356609v11i3p813

Routledge, P. (2012). Translocal climate justice solidarities. In *The Oxford handbook of climate change and society*. https://doi.org/10.1093/oxfordhb/9780199566600.003.0026

Schlosberg, D. (2012). Climate justice and capabilities: A framework for adaptation policy. *Ethics & International Affairs, 26*(4), 445–461.

Schlosberg, D., & Carruthers, D. (2010). Indigenous struggles, environmental justice, and community capabilities. *Global Environmental Politics, 10*(4), 12–35. https://doi.org/10.1162/GLEP_a_00029

Schlosberg, D., Rickards, L., & Byrne, J. (2018). Environmental justice and attachment to place. In R. Holifield, J. Chakraborty, & G. Walker (Eds.), *The Routledge handbook of environmental justice* (pp. 591–602). https://doi.org/10.4324/9781315678986-47

Sen, A. (1980). Equality of what? *Tanner Lectures on Human Values*, 197–220.

Sen, A. (1999). *Development as freedom*. Oxford: Oxford University Press. https://doi.org/10.1215/0961754X-9-2-350

Setzer, J., & Byrnes, R. (2020). *Global trends in climate change litigation: 2020 snapshot*. London: Grantham Research Institute on Climate Change and the Environment and Centre for Climate Change Economics and Policy, London School of Economics and Political Science.

Shaw, A. (2018). Environmental justice for a changing Arctic and its original peoples. In R. Holifield, J. Chakraborty, & G. Walker (Eds.), *The Routledge handbook of environmental justice* (pp. 504–514). https://doi.org/10.4324/9781315678986-40

United Nations Development Programme. (2016). *Tulele peisa, papua new guinea. Equator initiative case study series*. New York, NY: United Nations Development Programme. Retrieved from https://www.equatorinitiative.org/wp-content/uploads/2017/05/case_1473429470.pdf

Van Asselt, H. (2016). The role of non-state actors in reviewing ambition, implementation, and compliance under the Paris agreement. *Climate Law, 6*(1–2), 91–108. https://doi.org/10.1163/18786561-00601006

Verne, J. (2012). *Living translocality: Space, culture and economy in contemporary Swahili trade*. Stuttgart: Franz Steiner Verlag.

Whyte, K. (2017). Is it colonial déjà vu? Indigenous peoples and climate injustice. In *Humanities for the environment: Integrating knowledge, forging new constellations of practice*. London and New York: Routledge. https://doi.org/10.4324/9781315642659

Williamson, D. H. Z., Yu, E. X., Hunter, C. M., Kaufman, J. A., Komro, K., Jelks, N. O., . . . Kegler, M. C. (2020). A scoping review of capacity-building efforts to address environmental justice concerns. *International Journal of Environmental Research and Public Health, 17*(11). https://doi.org/10.3390/ijerph17113765

Zhang, Y.-X., Chao, Q.-C., Zheng, Q.-H., & Huang, L. (2017). The withdrawal of the U.S. from the Paris agreement and its impact on global climate change governance. *Advances in Climate Change Research, 8*(4), 213–219. https://doi.org/https://doi.org/10.1016/j.accre.2017.08.005

Index

Printed in the United States
by Baker & Taylor Publisher Services